Synthesis Lectures on Mathematics & Statistics

Series Editor

Steven G. Krantz, Department of Mathematics, Washington University, Saint Louis, MO, USA

This series includes titles in applied mathematics and statistics for cross-disciplinary STEM professionals, educators, researchers, and students. The series focuses on new and traditional techniques to develop mathematical knowledge and skills, an understanding of core mathematical reasoning, and the ability to utilize data in specific applications.

Mouffak Benchohra · Erdal Karapınar ·
Jamal Eddine Lazreg · Abdelkrim Salim

Advanced Topics
in Fractional Differential
Equations

A Fixed Point Approach

 Springer

Mouffak Benchohra
Laboratory of Mathematics
Djillali Liabes University
Sidi Bel-Abbes, Algeria

Erdal Karapınar
Department of Mathematics
Çankaya University
Ankara, Turkey

Jamal Eddine Lazreg
Laboratory of Mathematics
Djillali Liabes University
Sidi Bel-Abbes, Algeria

Abdelkrim Salim
Faculty of Technology
Hassiba Benbouali University
Chlef, Algeria

ISSN 1938-1743   ISSN 1938-1751 (electronic)
Synthesis Lectures on Mathematics & Statistics
ISBN 978-3-031-26930-1   ISBN 978-3-031-26928-8 (eBook)
https://doi.org/10.1007/978-3-031-26928-8

This Springer imprint is published by the registered company Springer Nature Switzerland AG
The registered company address is: Gewerbestrasse 11, 6330 Cham, Switzerland

We dedicate this book to our family members. In particular, Mouffak Benchohra makes his dedication to the memory of his father Yahia Benchohra and his wife Kheira Bencherif; Erdal Karapınar dedicates it to the memory of his father Hüseyin Karapınar and his mother Elife Karapınar; Jamal E. Lazreg dedicates it to the memory of his father Mohammed Lazreg; and Abdelkrim Salim makes his dedication to his mother, his brother, and his sisters.

Preface

Fractional Calculus is a branch of Mathematics which deals with the derivatives and integrals of arbitrary (non-integer) order. Though the subject is as old as conventional Calculus, the applications are rather recent. The researchers Grunwald, Letnikov, L'Hospital, Leibnitz, Hardy, Caputo, Mainardi, and others did pioneering work in this field. The key feature of the models involving fractional derivative is the flexibility in the choice of fractional order. Such models are proved appropriate in modeling the processes showing an intermediate behavior. The fractional order operators are nonlocal in contrast with the classical integer-order operator. This nonlocality plays a vital role in modeling the memory and hereditary properties in the natural systems.

This book is devoted to the existence, uniqueness, and stability results for various classes of problems with different conditions. All of the problems in this book deal with fractional differential equations and some form of extension of the well-known Hilfer fractional derivative which unifies the Riemann-Liouville and Caputo fractional derivatives. We made certain that each chapter contains results that may be regarded as a generalization or a partial continuation of the prior chapter's results. Classical and new fixed point theorems associated with the concept of measure of noncompactness in Banach spaces, as well as several generalizations of Gronwall's lemma, are employed as tools. Each chapter ends with a section devoted to remarks and bibliographical suggestions, and all abstract results are substantiated with illustrations.

This monograph adds to the current literature on fractional calculus by providing original content. All of the chapters include some of the authors' most current research work on the topic. This book is appropriate for use in advanced graduate courses, seminars, and research projects in numerous applied sciences.

We are grateful to J. R. Graef, J. Henderson, G. N'Guérékata, J. J. Nieto, and Y. Zhou for their contributions to research on the problems covered in this book.

Sidi Bel-Abbès, Algeria   Mouffak Benchohra
Ankara, Turkey   Erdal Karapınar
Sidi Bel-Abbès, Algeria   Jamal Eddine Lazreg
Sidi Bel-Abbès, Algeria   Abdelkrim Salim

Contents

Introduction

Fractional calculus is a field in mathematical analysis which is a generalization of integer differential calculus that involves real or complex order derivatives and integrals [10–14, 25, 28, 43, 50–52]. There is a long history of this concept of fractional differential calculus. One might wonder what meaning could be attributed to the derivative of a fractional order, that is $\frac{d^n y}{dx^n}$, where n is a fraction. Indeed, in correspondence with Leibniz, L'Hopital considered this very possibility. L'Hopital wrote to Leibniz in 1695 asking, "What if n be $\frac{1}{2}$?" The study of the fractional calculus was born from this question. Leibniz responded to the question, "$d^{\frac{1}{2}} x$ will be equal to $x\sqrt{dx : x}$. This is an apparent paradox from which, one day, useful consequences will be drawn."

Over the years, many well-known mathematicians have assisted in this theory. Thus, 30 September 1695 is the precise date of birth of the "fractional calculus"! Consequently, the fractional calculus has its roots in the work of Leibnitz, L'Hopital (1695), Bernoulli (1697), Euler (1730), and Lagrange (1772). Some years later, Laplace (1812), Fourier (1822), Abel (1823), Liouville (1832), Riemann (1847), Grünwald (1867), Letnikov (1868), Nekrasov (1888), Hadamard (1892), Heaviside (1892), Hardy (1915), Weyl (1917), Riesz (1922), P. Levy(1923), Davis (1924), Kober (1940), Zygmund (1945), Kuttner (1953), J. L. Lions (1959), Liverman (1964), and several more have developed the fundamental principle of fractional calculus.

Ross held the first fractional calculus conference at the University of New Haven in June of 1974 and edited its proceedings [118]. Thereafter, Spanier published the first monograph devoted to "Fractional Calculus" in 1974 [107]. In recent research in theoretical physics, mechanics, and applied mathematics; the integrals and derivatives of non-integer order; and the fractional integrodifferential equations have seen numerous applications. Samko, Kilbas, and Marichev's exceptionally detailed encyclopedic-type monograph was published in Russian in 1987 and in English in 1993 [138], (for more details, see [95]). The works devoted substantially to fractional differential equations are the book of Miller and Ross

M. Benchohra et al., *Advanced Topics in Fractional Differential Equations*,
Synthesis Lectures on Mathematics & Statistics,
https://doi.org/10.1007/978-3-031-26928-8_1

[99], of Podlubny [111], by Kilbas et al. [85], by Diethelm [67], by Ortigueira [108], by Abbas et al. [14], and by Baleanu et al. [42].

The origins of fixed point theory, as it is very well-known, go to the system of successive approximations (or the iterative method of Picard) used to solve certain differential equations. Roughly speaking, from the process of successive approximations, Banach obtained the fixed point theorem. The fixed point theory has been immense and independent of the differential equations in the last few decades. But, lately, the outcomes of fixed points have turned out to be the instruments for the differential equation's solutions. Recently, differential fractional order equations have been shown to be an effective instrument for researching multiple phenomena in diverse fields of science and engineering, such as electrochemistry, electromagnetics, viscoelasticity, and economics. It is very popular in the literature to suggest a solution to fractional differential equations by adding various forms of fractional derivatives; see, e.g., [7–9, 13, 14, 17, 19–21, 25, 28, 34, 43, 46, 58, 83, 84, 162]. In the other hand, there are more findings concerned with the issues of boundary value for fractional differential equations [25, 40, 48, 49, 59, 162].

In 1940, Ulam [152, 153] raised the following problem of the stability of the functional equation (of group homomorphisms): *"Under what conditions does it exist an additive mapping near an approximately additive mapping ?"*

Let G_1 be a group and let G_2 be a metric group with a metric $d(\cdot, \cdot)$. Given any $\epsilon > 0$, does there exist a $\delta > 0$ such that if a function $h : G_1 \to G_2$ satisfies the inequality $d(h(xy), h(x)h(y)) < \delta$ for all $x, y \in G_1$, then there exists a homomorphism $H : G_1 \to G_2$ with $d(h(x), H(x)) < \epsilon$ for all $x \in G_1$?

A partial answer was given by Hyers [80] in 1941, and between 1982 and 1998 Rassias [116, 117] established the Hyers-Ulam stability of linear and nonlinear mappings. Subsequently, many works have been published in order to generalize Hyers results in various directions; see, for example, [10, 13, 51, 52, 80, 89, 96, 114, 115, 119, 141, 153].

Many physical phenomena have short-term perturbations at some points caused by external interventions during their evolution. Adequate models for this kind of phenomenon are impulsive differential equations. Two types of impulses are popular in the literature: instantaneous impulses (whose duration is negligible) and non-instantaneous impulses (these changes start impulsively and remain active on finite initially given time intervals). There are mainly two approaches for the interpretation of the solutions of impulsive fractional differential equations: one by keeping the lower bound of the fractional derivative at the fixed initial time and the other by switching the lower limit of the fractional derivative at the impulsive points. The statement of the problem depends significantly on the type of fractional derivative. Fractional derivatives have some properties similar to ordinary derivatives (such as the derivative of a constant) which lead to similar initial value problems as well as similar impulsive conditions (instantaneous and non-instantaneous). The class of problems for fractional differential equations with abrupt and instantaneous impulses is vastly studied, and different topics on the existence and qualitative properties of solutions are considered, [50, 69, 154]. In pharmacotherapy, instantaneous impulses cannot describe the dynamics of certain evolution processes. For example, when one considers the hemodynamic equilibrium of a person, the introduction of the drugs in the bloodstream and the consequent absorption by the body are a gradual and continuous process. In the literature, many types of initial value problems and boundary value problems for different fractional differential equations with instantaneous and non-instantaneous impulses are studied (see, for example, [1, 3–8, 15, 24, 39, 50, 78, 87, 150, 155, 157]).

The measure of noncompactness, which is one of the fundamental tools in the theory of nonlinear analysis, was initiated by the pioneering articles of Alvàrez [35] and Mönch [101] and was developed by Banas and Goebel [45] and many researchers in the literature. The applications of the measure of noncompactness can be seen in the wide range of applied mathematics: theory of differential equations (see [22, 109] and references therein). Recently, in [13, 35, 38, 45], the authors applied the measure of noncompactness to some classes of differential equations in Banach spaces.

Nonlocal conditions were initiated by Byszewski [60] when he proved the existence and uniqueness of mild and classical solutions of nonlocal Cauchy problems. The nonlocal condition can be more useful than the standard initial condition to describe some physical phenomena. Fractional differential equations with nonlocal conditions have been discussed in [18, 26, 105] and references therein.

Many articles and monographs have been written recently in which the authors investigated numerous results for systems with different types of differential and integral equations and inclusions and various conditions. One may see the papers [16, 30, 36, 47, 74, 77, 92, 120, 121, 135, 136, 140] and the references therein.

One of the primary topics of this monograph is a new generalization of the well-known Hilfer fractional derivative, as well as a generalization of Grönwall's lemma and the many types of Ulam stability. In fact, this form of fractional derivative appears in the majority of the problems covered in this book. In order to define this new derivative, we took the publications of Diaz et al. [66] into account, where they presented the k-gamma and k-beta functions and demonstrated a number of their properties, many of which can also be found in [63, 102–104]. In addition, we were inspired by Sousa's numerous publications [143–149], in which they established another sort of fractional operator known as the ψ-Hilfer fractional derivative with respect to a particular function and provided several essential properties about this type of fractional operator. Our work on this monograph may be viewed as a continuation and generalization of the preceding studies, i.e., several results in the fractional calculus literature.

In the following, we give an outline of this monograph organization, which consists of five chapters defining the contributed work.

Chapter 2 provides the notation and preliminary results, descriptions, theorems, and other auxiliary results that will be needed for this study. In the first section, we give some notations and definitions of the functional spaces used in this book. In the second section, we give the definitions of the elements from fractional calculus theory, then we present some necessary lemmas, theorems, and properties. In the third section, we give some properties

to the measure of noncompactness. We finish the chapter in the last section by giving all the
fixed point theorems that are used throughout the book.

Chapter 3 deals with some existence and Ulam stability results for a class of initial and
boundary value problems for differential equations with generalized Hilfer-type fractional
derivative in Banach spaces. The chapter is divided into six sections. We start with Sect. 3.1,
which provides an introduction and some motivations, then finish the chapter with Sect. 3.6,
which contains some remarks and suggestions. The main results of the chapter begin with
Sect. 3.2; in it, we provide some existence results for the boundary value problem of the
following generalized Hilfer-type fractional differential equation:

$$\begin{cases} \left({}^\rho D_{a+}^{\alpha,\beta} u \right)(t) = f\left(t, u(t), \left({}^\rho D_{a+}^{\alpha,\beta} u \right)(t) \right), & t \in (a, b], \\ l\left({}^\rho J_{a+}^{1-\gamma} u \right)(a^+) + m\left({}^\rho J_{a+}^{1-\gamma} u \right)(b) = \phi, \end{cases}$$

where ${}^\rho D_{a+}^{\alpha,\beta}, {}^\rho J_{a+}^{1-\gamma}$ are the generalized Hilfer-type fractional derivative of order $\alpha \in (0, 1)$
and type $\beta \in [0, 1]$ and generalized fractional integral of order $1 - \gamma$, $(\gamma = \alpha + \beta - \alpha\beta)$,
respectively, $\phi \in E, 0 < a < b < +\infty$, $f : (a, b] \times E \times E \to E$ is a given function where
$(E, \| \cdot \|)$ is a Banach space and l, m are reals with $l + m \neq 0$. The results are based on
the fixed point theorems of Darbo and Mönch associated with the technique of measure of
noncompactness. Next, we prove that our problem is generalized Ulam-Hyers-Rassias stable.
An example is included to show the applicability of our results. In Sect. 3.3, we prove some
existence, uniqueness, and Ulam-Hyers-Rassias stability results for the following initial
value problem for implicit nonlinear fractional differential equations and k-generalized ψ-
Hilfer fractional derivative:

$$\begin{cases} \left({}^H_k D_{a+}^{\alpha,\beta;\psi} x \right)(t) = f\left(t, x(t), \left({}^H_k D_{a+}^{\alpha,\beta;\psi} x \right)(t) \right), & t \in (a, b], \\ \left(J_{a+}^{k(1-\xi),k;\psi} x \right)(a^+) = x_0, \end{cases}$$

where ${}^H_k D_{a+}^{\alpha,\beta;\psi}, J_{a+}^{k(1-\xi),k;\psi}$ are the k-generalized ψ-Hilfer fractional derivative of order
$\alpha \in (0, 1)$ and type $\beta \in [0, 1]$, and k-generalized ψ-fractional integral of order $k(1 - \xi)$,
where $\xi = \frac{1}{k}(\beta(k - \alpha) + \alpha), x_0 \in \mathbb{R}, k > 0$, and $f \in C([a, b] \times \mathbb{R}^2, \mathbb{R})$. The result is based
on the Banach contraction principle. In addition, two examples are given for justifying our
results. Section 3.4 deals with some existence and Ulam-Hyers-Rassias stability results for
the following initial value problem for implicit nonlinear fractional differential equations
and generalized ψ-Hilfer fractional derivative in Banach spaces:

$$\begin{cases} \left({}^H_k D_{a+}^{\alpha,\beta;\psi} x \right)(t) = f\left(t, x(t), \left({}^H_k D_{a+}^{\alpha,\beta;\psi} x \right)(t) \right), & t \in (a, b], \\ \left(J_{a+}^{k(1-\xi),k;\psi} x \right)(a^+) = x_0, \end{cases}$$

where $f \in C([a, b] \times E \times E, E)$. The results are based on fixed point theorems of Darbo and Mönch associated with the technique of measure of noncompactness. Illustrative examples are the subject of the last part. In Sect. 3.5, we prove some existence, uniqueness, and k-Mittag-Leffler-Ulam-Hyers stability results for the following boundary value problem for implicit nonlinear fractional differential equations and k-generalized ψ-Hilfer fractional derivative:

$$
\begin{cases}
\left({}^{H}_{k}\mathcal{D}^{\alpha,\beta;\psi}_{a+} x \right)(t) = f\left(t, x(t), \left({}^{H}_{k}\mathcal{D}^{\alpha,\beta;\psi}_{a+} x \right)(t) \right), & t \in (a, b], \\
c_1 \left(\mathcal{J}^{k(1-\xi),k;\psi}_{a+} x \right)(a^+) + c_2 \left(\mathcal{J}^{k(1-\xi),k;\psi}_{a+} x \right)(b) = c_3,
\end{cases}
$$

where $f \in C([a, b] \times \mathbb{R}^2, \mathbb{R})$ and $c_1, c_2, c_3 \in \mathbb{R}$ such that $c_1 + c_2 \neq 0$. Finally, several examples are given for justifying our results and addressing the different specific cases of our problem.

The aim of Chap. 4 is to prove some existence, uniqueness, and Ulam-Hyers-Rassias stability results for a class of boundary value problem for nonlinear implicit fractional differential equations with impulses and generalized Hilfer-type fractional derivative. We base our arguments on some relevant fixed point theorems combined with the technique of measure of noncompactness. Examples are included to show the applicability of our results for each section. The first result is provided in Sect. 4.2; in it, we establish existence, uniqueness, and Ulam-Hyers-Rassias results to the boundary value problem with nonlinear implicit generalized Hilfer-type fractional differential equation with impulses:

$$
\begin{cases}
\left({}^{\rho}\mathcal{D}^{\alpha,\beta}_{t_k^+} u \right)(t) = f\left(t, u(t), \left({}^{\rho}\mathcal{D}^{\alpha,\beta}_{t_k^+} u \right)(t) \right); & t \in J_k, \; k = 0, \ldots, m, \\
\left({}^{\rho}\mathcal{J}^{1-\gamma}_{t_k^+} u \right)(t_k^+) = \left({}^{\rho}\mathcal{J}^{1-\gamma}_{t_{k-1}^+} u \right)(t_k^-) + L_k(u(t_k^-)); & k = 1, \ldots, m, \\
c_1 \left({}^{\rho}\mathcal{J}^{1-\gamma}_{a+} u \right)(a^+) + c_2 \left({}^{\rho}\mathcal{J}^{1-\gamma}_{t_m^+} u \right)(b) = c_3,
\end{cases}
$$

where ${}^{\rho}\mathcal{D}^{\alpha,\beta}_{t_k^+}, {}^{\rho}\mathcal{J}^{1-\gamma}_{t_k^+}$ are the generalized Hilfer fractional derivative of order $\alpha \in (0, 1)$ and type $\beta \in [0, 1]$ and generalized fractional integral of order $1 - \gamma$, $(\gamma = \alpha + \beta - \alpha\beta)$, respectively, c_1, c_2, c_3 are reals with $c_1 + c_2 \neq 0$, $J_k := (t_k, t_{k+1}]$; $k = 0, \ldots, m$, $a = t_0 < t_1 < \cdots < t_m < t_{m+1} = b < \infty$, $u(t_k^+) = \lim_{\epsilon \to 0^+} u(t_k + \epsilon)$ and $u(t_k^-) = \lim_{\epsilon \to 0^-} u(t_k + \epsilon)$ represent the right- and left-hand limits of $u(t)$ at $t = t_k$, $f : (a, b] \times \mathbb{R} \times \mathbb{R} \to \mathbb{R}$ is a given function, and $L_k : \mathbb{R} \to \mathbb{R}$; $k = 1, \ldots, m$ are given continuous functions. The results are based on the Banach contraction principle and Krasnoselskii's and Schaefer's fixed point theorems. In Sect. 4.3, we examine the existence and the Ulam stability of the solutions to the boundary value problem with nonlinear implicit generalized Hilfer-type fractional differential equation with instantaneous impulses:

$$\begin{cases} \left({}^{\rho}\mathcal{D}_{t_k^+}^{\alpha,\beta}u\right)(t) = f\left(t, u(t), \left({}^{\rho}\mathcal{D}_{t_k^+}^{\alpha,\beta}u\right)(t)\right); \; t \in J_k, \; k = 0, \cdots, m, \\ \left({}^{\rho}\mathcal{J}_{t_k^+}^{1-\gamma}u\right)(t_k^+) = \left({}^{\rho}\mathcal{J}_{t_{k-1}^+}^{1-\gamma}u\right)(t_k^-) + \varpi_k(u(t_k^-)); \; k = 1, \cdots, m, \\ c_1\left({}^{\rho}\mathcal{J}_{a^+}^{1-\gamma}u\right)(a^+) + c_2\left({}^{\rho}\mathcal{J}_{t_m^+}^{1-\gamma}u\right)(b) = c_3, \end{cases}$$

where ${}^{\rho}\mathcal{D}_{t_k^+}^{\alpha,\beta}$, ${}^{\rho}\mathcal{J}_{t_k^+}^{1-\gamma}$ are the generalized Hilfer fractional derivative of order $\alpha \in (0,1)$ and type $\beta \in [0,1]$ and generalized Hilfer fractional integral of order $1-\gamma$, $(\gamma = \alpha + \beta - \alpha\beta)$, respectively, c_1, c_2 are reals with $c_1 + c_2 \neq 0$, $J_k := (t_k, t_{k+1}]; k = 0, \ldots, m$, $a = t_0 < t_1 < \cdots < t_m < t_{m+1} = b < \infty, u(t_k^+) = \lim_{\epsilon \to 0^+} u(t_k + \epsilon)$ and $u(t_k^-) = \lim_{\epsilon \to 0^-} u(t_k + \epsilon)$ represent the right- and left-hand limits of $u(t)$ at $t = t_k$, $c_3 \in E$, $f : (a,b] \times E \times E \to E$ is a given function, and $\varpi_k : E \to E; k = 1, \ldots, m$ are given continuous functions, where $(E, \|\cdot\|)$ is a Banach space. The results are based on fixed point theorems of Darbo and Mönch associated with the technique of measure of noncompactness. Examples are included to show the applicability of our results for each case.

Chapter 5 deals with some existence, uniqueness, and Ulam stability results for a class of initial and boundary value problems for nonlinear implicit fractional differential equations with non-instantaneous impulses and generalized Hilfer-type fractional derivative. The tools employed are some suitable fixed point theorems combined with the technique of measure of noncompactness. We provide illustrations to demonstrate the applicability of our results for each section. After the introduction section, in Sect. 5.2, we present some existence results to the initial value problem with nonlinear implicit generalized Hilfer-type fractional differential equation with non-instantaneous impulses:

$$\begin{cases} \left({}^{\rho}\mathcal{D}_{s_k^+}^{\alpha,\beta}u\right)(t) = f\left(t, u(t), \left({}^{\rho}\mathcal{D}_{s_k^+}^{\alpha,\beta}u\right)(t)\right); \; t \in I_k, \; k = 0, \ldots, m, \\ u(t) = g_k(t, u(t)); \; t \in \tilde{I}_k, \; k = 1, \ldots, m, \\ \left({}^{\rho}\mathcal{J}_{a^+}^{1-\gamma}u\right)(a^+) = \phi_0, \end{cases}$$

where ${}^{\rho}\mathcal{D}_{s_k^+}^{\alpha,\beta}$, ${}^{\rho}\mathcal{J}_{a^+}^{1-\gamma}$ are the generalized Hilfer fractional derivative of order $\alpha \in (0,1)$ and type $\beta \in [0,1]$ and generalized fractional integral of order $1-\gamma$, $(\gamma = \alpha + \beta - \alpha\beta)$, respectively, $\phi_0 \in \mathbb{R}$, $I_k := (s_k, t_{k+1}]; k = 0, \ldots, m$, $\tilde{I}_k := (t_k, s_k]; k = 1, \ldots, m, a = t_0 = s_0 < t_1 \leq s_1 < t_2 \leq s_2 < \cdots \leq s_{m-1} < t_m \leq s_m < t_{m+1} = b < \infty$, $u(t_k^+) = \lim_{\epsilon \to 0^+} u(t_k + \epsilon)$ and $u(t_k^-) = \lim_{\epsilon \to 0^-} u(t_k + \epsilon)$ represent the right- and left-hand limits of $u(t)$ at $t = t_k$, $f : (a,b] \times \mathbb{R} \times \mathbb{R} \to \mathbb{R}$ is a given function, and $g_k : \tilde{I}_k \times \mathbb{R} \to \mathbb{R}; k = 1, \ldots, m$, are given continuous functions such that $\left({}^{\rho}\mathcal{J}_{s_k^+}^{1-\gamma}g_k\right)(t, u(t))|_{t=s_k} = \phi_k \in \mathbb{R}$. The results are based on the Banach contraction principle and Schaefer's fixed point theorem. In Sect. 5.2.2, we give a generalization of the previous result to nonlocal impulsive fractional differential equations. More precisely, we present some existence results for the following nonlocal problem:

$$
\begin{cases}
\left({}^{\rho}\mathcal{D}^{\alpha,\beta}_{s_k^+} u \right)(t) = f\left(t, u(t), \left({}^{\rho}\mathcal{D}^{\alpha,\beta}_{s_k^+} u \right)(t) \right); \ t \in I_k, \ k = 0, \ldots, m, \\
u(t) = g_k(t, u(t)); \ t \in \tilde{I}_k, \ k = 1, \ldots, m, \\
\left({}^{\rho}\mathcal{J}^{1-\gamma}_{a^+} u \right)(a^+) + \xi(u) = \phi_0,
\end{cases}
$$

where ξ is a continuous function. In Sect. 5.3, we establish some existence results to the initial value problem of nonlinear implicit generalized Hilfer-type fractional differential equation with non-instantaneous impulses:

$$
\begin{cases}
\left({}^{\rho}\mathcal{D}^{\alpha,\beta}_{s_k^+} u \right)(t) = f\left(t, u(t), \left({}^{\rho}\mathcal{D}^{\alpha,\beta}_{s_k^+} u \right)(t) \right); \ t \in I_k, \ k = 0, \ldots, m, \\
u(t) = g_k(t, u(t)); \ t \in \tilde{I}_k, \ k = 1, \ldots, m, \\
\\
\left({}^{\rho}\mathcal{J}^{1-\gamma}_{a^+} u \right)(a^+) = \phi_0,
\end{cases}
$$

where ${}^{\rho}\mathcal{D}^{\alpha,\beta}_{s_k^+}$, ${}^{\rho}\mathcal{J}^{1-\gamma}_{a^+}$ are the generalized Hilfer-type fractional derivative of order $\alpha \in (0, 1)$ and type $\beta \in [0, 1]$ and generalized fractional integral of order $1 - \gamma$, $(\gamma = \alpha + \beta - \alpha\beta)$, respectively, $\rho > 0$, $\phi_0 \in E$, $I_k := (s_k, t_{k+1}]; k = 0, \ldots, m$, $\tilde{I}_k := (t_k, s_k]; k = 1, \ldots, m$, $a = s_0 < t_1 \leq s_1 < t_2 \leq s_2 < \cdots \leq s_{m-1} < t_m \leq s_m < t_{m+1} = b < \infty$, $u(t_k^+) = \lim_{\epsilon \to 0^+} u(t_k + \epsilon)$ and $u(t_k^-) = \lim_{\epsilon \to 0^-} u(t_k + \epsilon)$ represent the right- and left-hand limits of $u(t)$ at $t = t_k$, $f : I_k \times E \times E \to E$ is a given function, and $g_k : \tilde{I}_k \times E \to E; k = 1, \ldots, m$ are given continuous functions such that $\left({}^{\rho}\mathcal{J}^{1-\gamma}_{s_k^+} g_k \right)(t, u(t)) \big|_{t=s_k} = \phi_k \in E$, where $(E, \|\cdot\|)$ is a real Banach space. The results are based on fixed point theorems of Darbo and Mönch associated with the technique of measure of noncompactness. Examples are included to show the applicability of our results. Section 5.4 presents some existence and stability results to the boundary value problem with nonlinear implicit generalized Hilfer-type fractional differential equation with non-instantaneous impulses:

$$
\begin{cases}
\left({}^{\alpha}\mathcal{D}^{\alpha,\beta}_{\tau_i^+} x \right)(t) = f\left(t, x(t), \left({}^{\alpha}\mathcal{D}^{\alpha,\beta}_{\tau_i^+} x \right)(t) \right); \ t \in J_i, \ i = 0, \ldots, m, \\
x(t) = \psi_i(t, x(t)); \ t \in \tilde{J}_i, \ i = 1, \ldots, m, \\
\phi_1 \left({}^{\alpha}\mathcal{J}^{1-\gamma}_{a^+} x \right)(a^+) + \phi_2 \left({}^{\alpha}\mathcal{J}^{1-\gamma}_{m^+} x \right)(b) = \phi_3,
\end{cases}
$$

where ${}^{\alpha}\mathcal{D}^{\alpha,\beta}_{\tau_i^+}, {}^{\alpha}\mathcal{J}^{1-\gamma}_{a^+}$ are the generalized Hilfer fractional derivative of order $\alpha \in (0, 1)$ and type $\beta \in [0, 1]$ and generalized fractional integral of order $1 - \gamma$, $(\gamma = \alpha + \beta - \alpha\beta)$, respectively, $\phi_1, \phi_2, \phi_3 \in \mathbb{R}$, $\phi_1 \neq 0$, $J_i := (\tau_i, t_{i+1}]; i = 0, \ldots, m$, $\tilde{J}_i := (t_i, s_i]; i = 1, \ldots, m$, $a = t_0 = \tau_0 < t_1 \leq \tau_1 < t_2 \leq \tau_2 < \cdots \leq \tau_{m-1} < t_m \leq \tau_m < t_{m+1} = b < \infty$,

$x(t_i^+) = \lim\limits_{\epsilon \to 0^+} x(t_i + \epsilon)$ and $x(t_i^-) = \lim\limits_{\epsilon \to 0^-} x(t_i + \epsilon)$ represent the right- and left-hand limits

of $x(t)$ at $t = t_i$, $f : (a, b] \times \mathbb{R} \times \mathbb{R} \to \mathbb{R}$ is a given function, and $\psi_i : \tilde{J}_i \times \mathbb{R} \to \mathbb{R}$; $i =$

$1, \ldots, m$ are given continuous functions such that $\left({}^{\alpha}\mathcal{J}_{\tau_i^+}^{1-\gamma} \psi_i \right) (t, x(t)) \big|_{t=\tau_i} = c_i \in \mathbb{R}$.

The results are based on the Banach contraction principle and Krasnoselskii's fixed point theorem. Further, for the justification of our results, we provide two examples.

Preliminary Background

In this chapter, we discuss the necessary mathematical tools, notations, and concepts we need in the succeeding chapters. We look at some essential properties of fractional differential operators. We also review some of the basic properties of measures of noncompactness and fixed point theorems which are crucial in our results regarding fractional differential equations.

2.1 Notations and Functional Spaces

In this section, we will provide all the notations and definitions of the functional spaces that are considered as fundamental and fixed throughout all the preceding chapters. Indeed, these are mentioned only one time in this section.

Let $0 < a < b$, $J = (a, b]$ where $\bar{J} = [a, b]$. Consider the following parameters α, β, γ satisfying $\gamma = \alpha + \beta - \alpha\beta$ and $0 < \alpha, \beta, \gamma < 1$. Let $\xi = \frac{1}{k}(\beta(k - \alpha) + \alpha)$ where $k > 0$. Let $\rho > 0$.

Let ψ be an increasing and positive function on $\bar{J}$ such that ψ' is continuous on $\bar{J}$.

2.1.1 Space of Continuous Functions

By $C(\bar{J}, \mathbb{R})$ we denote the Banach space of all continuous functions from $\bar{J}$ into $\mathbb{R}$ with the norm

$$\|u\|_\infty = \sup\{|u(t)| : t \in \bar{J}\}.$$

Let $(E, \|\cdot\|)$ be a Banach space. By $C(\bar{J}, E)$ we denote the Banach space of all continuous functions from $\bar{J}$ into E with the norm

M. Benchohra et al., *Advanced Topics in Fractional Differential Equations*,
Synthesis Lectures on Mathematics & Statistics,
https://doi.org/10.1007/978-3-031-26928-8_2

$$\|u\|_E = \sup\{\|u(t)\| : t \in \bar{J}\}.$$

$AC^n(J, \mathbb{R})$, $C^n(J, \mathbb{R})$ are the spaces of n-times absolutely continuous and n-times continuously differentiable functions on J, respectively.

2.1.2 Spaces of Integrable Functions

Consider the space $X_c^p(a, b)$, $(c \in \mathbb{R}, \ 1 \le p \le \infty)$ of those complex-valued Lebesgue measurable functions f on $\bar{J}$ for which $\|f\|_{X_c^p} < \infty$, where the norm is defined by

$$\|f\|_{X_c^p} = \left(\int_a^b |t^c f(t)|^p \frac{dt}{t} \right)^{\frac{1}{p}}, \quad (1 \le p < \infty, c \in \mathbb{R}).$$

In particular, when $c = \frac{1}{p}$, the space $X_c^p(a, b)$ coincides with the $L^p(a, b)$ space: $X_{\frac{1}{p}}^p(a, b) = L^p(a, b)$.

By $L^1(J)$, we denote the space of Bochner-integrable functions $f : J \longrightarrow E$ with the norm

$$\|f\|_1 = \int_a^b \|f(t)\| dt.$$

Consider the space $X_\psi^p(a, b)$, $(1 \le p \le \infty)$ of those real-valued Lebesgue measurable functions g on $\bar{J}$ for which $\|g\|_{X_\psi^p} < \infty$, where the norm is defined by

$$\|g\|_{X_\psi^p} = \left(\int_a^b \psi'(t) |g(t)|^p dt \right)^{\frac{1}{p}},$$

where ψ is an increasing and positive function on $[a, b]$ such that ψ' is continuous on $\bar{J}$. In particular, when $\psi(x) = x$, the space $X_\psi^p(a, b)$ coincides with the $L_p(a, b)$ space.

2.1.3 Spaces of Continuous Functions with Weight

We consider the weighted spaces of continuous functions

$$C_{\gamma, \rho}(J) = \left\{ u : J \to E : \left(\frac{t^\rho - a^\rho}{\rho} \right)^{1-\gamma} u(t) \in C\left(\bar{J}, E\right) \right\},$$

and

$$C_{\gamma, \rho}^n(J) = \left\{ u \in C^{n-1} : u^{(n)} \in C_{\gamma, \rho}(J) \right\}, n \in \mathbb{N},$$
$$C_{\gamma, \rho}^0(J) = C_{\gamma, \rho}(J),$$

with the norms

$$\|u\|_{C_{\gamma,\rho}} = \sup_{t \in \bar{J}} \left\| \left(\frac{t^\rho - a^\rho}{\rho} \right)^{1-\gamma} u(t) \right\|,$$

and

$$\|u\|_{C^n_{\gamma,\rho}} = \sum_{k=0}^{n-1} \|u^{(k)}\|_\infty + \|u^{(n)}\|_{C_{\gamma,\rho}}.$$

We define the spaces

$$C^{\alpha,\beta}_{\gamma,\rho}(J) = \left\{ u \in C_{\gamma,\rho}(J),\ {}^\rho D^{\alpha,\beta}_{a^+} u \in C_{\gamma,\rho}(J) \right\},$$

and

$$C^\gamma_{\gamma,\rho}(J) = \left\{ u \in C_{\gamma,\rho}(J),\ {}^\rho D^\gamma_{a^+} u \in C_{\gamma,\rho}(J) \right\},$$

where ${}^\rho D^{\alpha,\beta}_{a^+}$ and ${}^\rho D^\gamma_{a^+}$ are fractional derivatives defined in the following sections.
    Consider the weighted Banach space

$$C_{\gamma;\psi}(J) = \left\{ u : J \to \mathbb{R} : t \to (\psi(t) - \psi(a))^{1-\gamma} u(t) \in C(\bar{J}, \mathbb{R}) \right\},$$

with the norm

$$\|u\|_{C_{\gamma;\psi}} = \sup_{t \in \bar{J}} \left| (\psi(t) - \psi(a))^{1-\gamma} u(t) \right|,$$

and

$$C^n_{\gamma;\psi}(J) = \left\{ u \in C^{n-1}(J) : u^{(n)} \in C_{\gamma;\psi}(J) \right\}, n \in \mathbb{N},$$

$$C^0_{\gamma;\psi}(J) = C_{\gamma;\psi}(J),$$

with the norm

$$\|u\|_{C^n_{\gamma;\psi}} = \sum_{i=0}^{n-1} \|u^{(i)}\|_\infty + \|u^{(n)}\|_{C_{\gamma;\psi}}.$$

The weighted space $C^{\alpha,\beta}_{\gamma;\psi}(J)$ is defined by

$$C^{\alpha,\beta}_{\gamma;\psi}(J) = \left\{ u \in C_{\gamma;\psi}(J),\ {}^H D^{\alpha,\beta;\psi}_{a^+} u \in C_{\gamma;\psi}(J) \right\}$$

where ${}^H D^{\alpha,\beta;\psi}_{a^+}$ is a fractional derivative defined in the following sections.
    Consider the weighted Banach space

$$C_{\xi;\psi}(J) = \left\{ x : J \to E : t \to (\psi(t) - \psi(a))^{1-\xi} x(t) \in C(\bar{J}, E) \right\},$$

with the norm

$$\|x\|_{C_{\xi;\psi}} = \sup_{t \in \bar{J}} \left\| (\psi(t) - \psi(a))^{1-\xi} x(t) \right\|,$$

and

$$C_{\xi;\psi}^n(J) = \left\{ x \in C^{n-1}(J) : x^{(n)} \in C_{\xi;\psi}(J) \right\}, n \in \mathbb{N},$$
$$C_{\xi;\psi}^0(J) = C_{\xi;\psi}(J),$$

with the norm

$$\|x\|_{C_{\xi;\psi}^n} = \sum_{i=0}^{n-1} \|x^{(i)}\|_\infty + \|x^{(n)}\|_{C_{\xi;\psi}}.$$

The weighted space $C_{\xi,k;\psi}^{\alpha,\beta}(J)$ is defined by

$$C_{\xi;\psi}^{\alpha,\beta}(J) = \left\{ x \in C_{\xi;\psi}(J), {}_k^H \mathcal{D}_{a+}^{\alpha,\beta;\psi} x \in C_{\xi;\psi}(J) \right\},$$

where ${}_k^H \mathcal{D}_{a+}^{\alpha,\beta;\psi}$ is defined in the sequel.

2.2 Special Functions of the Fractional Calculus

2.2.1 Gamma Function

Undoubtedly, one of the basic functions of the fractional calculus is Euler's gamma function $\Gamma(z)$, which generalizes the factorial $n!$ and allows n to take also non-integer and even complex values. Leonhard Euler was a Swiss mathematician, physicist, astronomer, geographer, logician, and engineer who pioneered and inspired discoveries in a wide range of mathematical fields, including analytic number theory, complex analysis, and infinitesimal calculus. He pioneered much of today's mathematical terminology and notation, including the notion of a mathematical function. In addition, he is well-known for his contributions to mechanics, fluid dynamics, optics, astronomy, and music theory.

Daniel Bernoulli then developed the gamma function for complex numbers with a positive real part. Daniel Bernoulli was a Swiss mathematician and physicist who was a member of the famous Bernoulli family from Basel. He is most known for his mathematical applications to mechanics, notably fluid mechanics, as well as his groundbreaking work in probability and statistics.

Definition 2.1 ([111]) The gamma function is defined via a convergent improper integral:

$$\Gamma(z) = \int_0^{+\infty} t^{z-1} e^{-t} \mathrm{dt},$$

where $z > 0$.

One of the basic properties of the gamma function is that it satisfies the following functional equation:

$$\Gamma(z+1) = z\Gamma(z),$$

so, for positive integer values n, the gamma function becomes $\Gamma(n) = (n-1)!$ and thus can be seen as an extension of the factorial function to real values. A useful particular value of the function: $\Gamma(\frac{1}{2}) = \sqrt{\pi}$ is used throughout many examples in this monograph.

2.2.2 k-Gamma and k-Beta Functions

In 2005, Diaz and Petruel [66] have defined new functions called k-gamma and k-beta functions given by

$$\Gamma_k(\alpha) = \int_0^\infty t^{\alpha-1} e^{-\frac{t^k}{k}} dt, \alpha > 0, k > 0$$

and

$$B_k(\alpha, \beta) = \frac{1}{k} \int_0^1 t^{\frac{\alpha}{k}-1}(1-t)^{\frac{\beta}{k}-1} dt.$$

It is noteworthy that if $k \to 1$ then $\Gamma_k(\alpha) \to \Gamma(\alpha)$ and $B_k(\alpha, \beta) \to B(\alpha, \beta)$. We have also the following useful relations:

$$\Gamma_k(\alpha) = k^{\frac{\alpha}{k}-1}\Gamma\left(\frac{\alpha}{k}\right), \quad \Gamma_k(\alpha+k) = \alpha\Gamma_k(\alpha), \quad \Gamma_k(k) = \Gamma(1) = 1,$$

$$B_k(\alpha, \beta) = \frac{1}{k} B\left(\frac{\alpha}{k}, \frac{\beta}{k}\right), \quad B_k(\alpha, \beta) = \frac{\Gamma_k(\alpha)\Gamma_k(\beta)}{\Gamma_k(\alpha+\beta)}.$$

2.2.3   Mittag-Leffler Function

The exponential function e^z plays a very important role in the theory of integer-order differential equations.

Definition 2.2 ([111]) The one-parameter generalization of the exponential function is now denoted by

$$\mathbb{E}_\alpha(z) = \sum_{k=0}^\infty \frac{z^k}{\Gamma(\alpha k + 1)}, \quad \alpha > 0.$$

Definition 2.3 ([111]) A two-parameter function of the Mittag-Leffler type is defined by the series expansion

$$\mathbb{E}_{\alpha,\beta}(z) = \sum_{k=0}^\infty \frac{z^k}{\Gamma(\alpha k + \beta)}, \quad \alpha > 0, \quad \beta > 0.$$

It follows from the definition that

$$\mathbb{E}_{1,1}(z) = \sum_{k=0}^{\infty} \frac{z^k}{\Gamma(k+1)} = \sum_{k=0}^{\infty} \frac{z^k}{k!} = e^z.$$

Definition 2.4 ([102]) The Mittag-Leffler function can also be refined into the k-Mittag-Leffler function defined as follows:

$$\mathbb{E}_k^{\alpha,\beta}(z) = \sum_{i=0}^{\infty} \frac{z^i}{\Gamma_k(\alpha i + \beta)}, \alpha, \beta > 0.$$

In this monograph, we will employ the following notation:

$$\mathbb{E}_k^{\alpha}(z) = \mathbb{E}_k^{\alpha,k}(z) = \sum_{i=0}^{\infty} \frac{z^i}{\Gamma_k(\alpha i + k)}, \alpha > 0.$$

2.3   Elements from Fractional Calculus Theory

In this section, we recall some definitions of fractional integral and fractional differential operators that include all we use throughout this monograph. We conclude it with some necessary lemmas, theorems, and properties.

2.3.1   Fractional Integrals

Definition 2.5 *(Generalized fractional integral* [85]) Let $\alpha \in \mathbb{R}_+$ and $g \in L^1(J)$. The generalized fractional integral of order α is defined by

$$\left({}^{\rho}\mathcal{J}_{a+}^{\alpha} g \right)(t) = \int_a^t s^{\rho-1} \left(\frac{t^{\rho} - s^{\rho}}{\rho} \right)^{\alpha-1} \frac{g(s)}{\Gamma(\alpha)} ds, \ t > a, \rho > 0.$$

Definition 2.6 *(ψ-Riemann-Liouville fractional integral* [85]) Let (a,b) $(-\infty \le a < b \le \infty)$ be a finite or infinite interval of the real line $\mathbb{R}$, $\alpha > 0$, $c \in \mathbb{R}$, and $h \in X_c^p(a,b)$. Also, let $\psi(t)$ be an increasing and positive monotone function on J, having a continuous derivative $\psi'(t)$ on (a,b). The left- and right-sided fractional integrals of a function h of order α with respect to another function ψ on J are defined by

$$\left(\mathcal{J}_{a+}^{\alpha;\psi} h \right)(t) = \int_a^t \psi'(\tau) (\psi(t) - \psi(\tau))^{\alpha-1} \frac{h(\tau)}{\Gamma(\alpha)} d\tau,$$

and

$$\left(\mathcal{J}_{b-}^{\alpha;\psi}h\right)(t) = \int_{t}^{b} \psi'(\tau)\left(\psi(\tau) - \psi(t)\right)^{\alpha-1}\frac{h(\tau)}{\Gamma(\alpha)}d\tau.$$

Definition 2.7 (*k-Generalized ψ-fractional integral* [113]) Let $g \in X_{\psi}^{p}(a, b)$, $\psi(t) > 0$ be an increasing function on J and $\psi'(t) > 0$ be continuous on (a, b) and $\alpha > 0$. The generalized k-fractional integral operators of a function g (left-sided and right-sided) of order α are defined by

$$\mathcal{J}_{a+}^{\alpha,k;\psi}g(t) = \frac{1}{k\Gamma_k(\alpha)}\int_{a}^{t}\frac{\psi'(s)g(s)ds}{(\psi(t) - \psi(s))^{1-\frac{\alpha}{k}}},$$

$$\mathcal{J}_{b-}^{\alpha,k;\psi}g(t) = \frac{1}{k\Gamma_k(\alpha)}\int_{t}^{b}\frac{\psi'(s)g(s)ds}{(\psi(s) - \psi(t))^{1-\frac{\alpha}{k}}},$$

with $k > 0$. Also, in [103], Nápoles Valdés gave more generalized fractional integral operators defined by

$$\mathcal{J}_{G,a+}^{\alpha,k;\psi}g(t) = \frac{1}{k\Gamma_k(\alpha)}\int_{a}^{t}\frac{\psi'(s)g(s)ds}{G(\psi(t) - \psi(s), \frac{\alpha}{k})},$$

$$\mathcal{J}_{G,b-}^{\alpha,k;\psi}g(t) = \frac{1}{k\Gamma_k(\alpha)}\int_{t}^{b}\frac{\psi'(s)g(s)ds}{G(\psi(s) - \psi(t), \frac{\alpha}{k})},$$

where $G(\cdot, \alpha) \in AC(J)$.

2.3.2   Fractional Derivatives

Definition 2.8 (*Generalized fractional derivative* [85]) Let $\alpha \in \mathbb{R}_+ \setminus \mathbb{N}$ and $\rho > 0$. The generalized fractional derivative $^{\rho}\mathcal{D}_{a+}^{\alpha}$ of order α is defined by

$$\left(^{\rho}\mathcal{D}_{a+}^{\alpha}g\right)(t) = \delta_{\rho}^{n}\left(^{\rho}\mathcal{J}_{a+}^{n-\alpha}g\right)(t)$$

$$= \left(t^{1-\rho}\frac{d}{dt}\right)^{n}\int_{a}^{t}s^{\rho-1}\left(\frac{t^{\rho} - s^{\rho}}{\rho}\right)^{n-\alpha-1}\frac{g(s)}{\Gamma(n-\alpha)}ds, \ t > a, \rho > 0,$$

where $n = [\alpha] + 1$ and $\delta_{\rho}^{n} = \left(t^{1-\rho}\frac{d}{dt}\right)^{n}$.

Definition 2.9 (*Generalized Hilfer-type fractional derivative* [106]) Let order α and type β satisfy $n - 1 < \alpha < n$ and $0 \leq \beta \leq 1$, with $n \in \mathbb{N}$. The generalized Hilfer-type fractional derivative to t, with $\rho > 0$ of a function g, is defined by

$$\left(^{\rho}\mathcal{D}_{a+}^{\alpha,\beta}g\right)(t) = \left(^{\rho}\mathcal{J}_{a+}^{\beta(n-\alpha)}\left(t^{\rho-1}\frac{d}{dt}\right)^{n}\ ^{\rho}\mathcal{J}_{a+}^{(1-\beta)(n-\alpha)}g\right)(t)$$

$$= \left(^{\rho}\mathcal{J}_{a+}^{\beta(n-\alpha)}\delta_{\rho}^{n}\ ^{\rho}\mathcal{J}_{a+}^{(1-\beta)(n-\alpha)}g\right)(t).$$

Definition 2.10 (ψ-*Riemann-Liouville fractional derivative* [85]) Let $\psi'(t) \neq 0$ ($-\infty \leq a < t < b \leq \infty$), $\alpha > 0$, and $n \in \mathbb{N}$. The Riemann-Liouville derivatives of a function h of order α with respect to another function ψ on $\bar{J}$ are defined by

$$\left(\mathcal{D}_{a+}^{\alpha;\psi} h\right)(t) = \delta^n (\mathcal{J}_{a+}^{n-\alpha;\psi} h)(t)$$

$$= \delta^n \int_a^t \psi'(\tau)\,(\psi(t) - \psi(\tau))^{n-\alpha-1} \frac{h(\tau)}{\Gamma(n-\alpha)} d\tau,$$

and

$$\left(\mathcal{D}_{b-}^{\alpha;\psi} h\right)(t) = (-1)^n \delta^n (\mathcal{J}_{a+}^{n-\alpha;\psi} h)(t)$$

$$= (-1)^n \delta^n \int_t^b \psi'(\tau)\,(\psi(\tau) - \psi(t))^{n-\alpha-1} \frac{h(\tau)}{\Gamma(n-\alpha)} d\tau,$$

where $n = [\alpha] + 1$ and $\delta^n = \left(\dfrac{1}{\psi'(t)}\dfrac{d}{dt}\right)^n$.

Definition 2.11 (ψ-*Hilfer fractional derivative* [148]) Let order α and type β satisfy $n - 1 < \alpha < n$ and $0 \leq \beta \leq 1$, with $n \in \mathbb{N}$, let $h, \psi \in C^n(\bar{J}, \mathbb{R})$ be two functions such that ψ is increasing and $\psi'(t) \neq 0$. The ψ-Hilfer fractional derivatives to t of a function h are defined by

$$\left(^H\mathcal{D}_{a+}^{\alpha,\beta;\psi} h\right)(t) = \left(\mathcal{J}_{a+}^{\beta(n-\alpha);\psi} \left(\frac{1}{\psi'(t)}\frac{d}{dt}\right)^n \mathcal{J}_{a+}^{(1-\beta)(n-\alpha);\psi} h\right)(t)$$

and

$$\left(^H\mathcal{D}_{b-}^{\alpha,\beta;\psi} h\right)(t) = \left(\mathcal{J}_{b-}^{\beta(n-\alpha);\psi} \left(-\frac{1}{\psi'(t)}\frac{d}{dt}\right)^n \mathcal{J}_{b-}^{(1-\beta)(n-\alpha);\psi} h\right)(t).$$

In this monograph, we consider the case $n = 1$ only, because $0 < \alpha < 1$.

We are now able to define the k-generalized ψ-Hilfer derivative as follows.

Definition 2.12 (k-*Generalized* ψ-*Hilfer derivative*) Let $n - 1 < \dfrac{\alpha}{k} \leq n$ with $n \in \mathbb{N}$, $-\infty \leq a < b \leq \infty$ and $g, \psi \in C^n(\bar{J}, \mathbb{R})$ be two functions such that ψ is increasing and $\psi'(t) \neq 0$, for all $t \in J$. The k-generalized ψ-Hilfer fractional derivatives (left-sided and right-sided) $^H_k\mathcal{D}_{a+}^{\alpha,\beta;\psi}(\cdot)$ and $^H_k\mathcal{D}_{b-}^{\alpha,\beta;\psi}(\cdot)$ of a function g of order α and type $0 \leq \beta \leq 1$, with $k > 0$, are defined by

$$^H_k\mathcal{D}_{a+}^{\alpha,\beta;\psi} g\,(t) = \left(\mathcal{J}_{a+}^{\beta(kn-\alpha),k;\psi} \left(\frac{1}{\psi'(t)}\frac{d}{dt}\right)^n \left(k^n\,\mathcal{J}_{a+}^{(1-\beta)(kn-\alpha),k;\psi} g\right)\right)(t)$$

$$= \left(\mathcal{J}_{a+}^{\beta(kn-\alpha),k;\psi} \delta_\psi^n \left(k^n\,\mathcal{J}_{a+}^{(1-\beta)(kn-\alpha),k;\psi} g\right)\right)(t),$$

and

$$
{}^{H}_{k}\mathcal{D}^{\alpha,\beta;\psi}_{b-}g\,(t)=\left(\mathcal{J}^{\beta(kn-\alpha),k;\psi}_{b-}\left(-\frac{1}{\psi'(t)}\frac{d}{dt}\right)^{n}\left(k^{n}\,\mathcal{J}^{(1-\beta)(kn-\alpha),k;\psi}_{b-}g\right)\right)(t)
$$
$$
=\left(\mathcal{J}^{\beta(kn-\alpha),k;\psi}_{b-}(-1)^{n}\delta^{n}_{\psi}\left(k^{n}\,\mathcal{J}^{(1-\beta)(kn-\alpha),k;\psi}_{b-}g\right)\right)(t),
$$

where $\delta^{n}_{\psi}=\left(\dfrac{1}{\psi'(t)}\dfrac{d}{dt}\right)^{n}$.

Property 2.13 It is worth noting that the k-generalized ψ-Hilfer fractional derivative is thought to be an expansion to many fractional operators defined over the years; indeed, in the following part, we will give a list of some of the most commonly used fractional derivatives that are considered to be a particular case of our operator. The fractional derivative ${}^{H}_{k}\mathcal{D}^{\alpha,\beta;\psi}_{a+}$ interpolates the following fractional derivatives:

- The ψ-Hilfer fractional derivative $(k=1)$;
- The ψ-Riemann-Liouville fractional derivative $(k=1,\beta=0)$;
- The ψ-Caputo fractional derivative $(k=1,\beta=1)$;
- The Hilfer fractional derivative $(k=1,\psi(t)=t)$;
- The Riemann-Liouville fractional derivative $(k=1,\psi(t)=t,\beta=0)$;
- The Caputo fractional derivative $(k=1,\psi(t)=t,\beta=1)$;
- The Hilfer-Hadamard fractional derivative $(k=1,\psi(t)=\ln(t))$;
- The Caputo-Hadamard fractional derivative $(k=1,\psi(t)=\ln(t),\beta=1)$;
- The Hadamard fractional derivative $(k=1,\psi(t)=\ln(t),\beta=0)$;
- The Hilfer-generalized fractional derivative $(k=1,\psi(t)=t^{\rho})$;
- The Caputo-generalized fractional derivative $(k=1,\psi(t)=t^{\rho},\beta=1)$;
- The generalized fractional derivative $(k=1,\psi(t)=t^{\rho},\beta=0)$;
- The Weyl fractional derivative $(k=1,\psi(t)=t^{\rho},\beta=0,a=-\infty)$.

2.3.3 Necessary Lemmas, Theorems, and Properties

Theorem 2.14 ([85]) Let $\alpha>0,\beta>0,0\leq\rho\leq\infty,0<a<b<\infty$. Then, for $g\in L^{1}(J)$ we have
$$
\left({}^{\rho}\mathcal{J}^{\alpha}_{a+}\,{}^{\rho}\mathcal{J}^{\beta}_{a+}g\right)(t)=\left({}^{\rho}\mathcal{J}^{\alpha+\beta}_{a+}g\right)(t).
$$

Lemma 2.15 ([145]) Let $\alpha>0,\ 0\leq\gamma<1$. Then, $\mathcal{J}^{\alpha;\psi}_{a+}$ is bounded from $C_{\gamma;\psi}(J)$ into $C_{\gamma;\psi}(J)$. In addition, if $\gamma\leq\alpha$, then $\mathcal{J}^{\alpha;\psi}_{a+}$ is bounded from $C_{\gamma;\psi}(J)$ into $C(\bar{J},\mathbb{R})$.

Theorem 2.16 ([103]) Let $g:\bar{J}\to\mathbb{R}$ be an integrable function, and take $\alpha>0$ and $k>0$. Then $\mathcal{J}^{\alpha,k;\psi}_{G,a+}g$ exists for all $t\in\bar{J}$.

Theorem 2.17 ([103]) *Let $g \in X^p_\psi(a, b)$ and take $\alpha > 0$ and $k > 0$. Then $\mathcal{J}^{\alpha,k;\psi}_{G,a+} g \in C(\bar{J}, \mathbb{R})$.*

Lemma 2.18 *Let $\alpha > 0$, $\beta > 0$, and $k > 0$. Then, we have the following semigroup property given by*

$$\mathcal{J}^{\alpha,k;\psi}_{a+} \mathcal{J}^{\beta,k;\psi}_{a+} f(t) = \mathcal{J}^{\alpha+\beta,k;\psi}_{a+} f(t) = \mathcal{J}^{\beta,k;\psi}_{a+} \mathcal{J}^{\alpha,k;\psi}_{a+} f(t),$$

and

$$\mathcal{J}^{\alpha,k;\psi}_{b-} \mathcal{J}^{\beta,k;\psi}_{b-} f(t) = \mathcal{J}^{\alpha+\beta,k;\psi}_{b-} f(t) = \mathcal{J}^{\beta,k;\psi}_{b-} \mathcal{J}^{\alpha,k;\psi}_{b-} f(t).$$

Proof By Lemma 1 in [148] and the property of k-gamma function, for $\alpha > 0$, $\beta > 0$, and $k > 0$, we get

$$\mathcal{J}^{\alpha,k;\psi}_{a+} \mathcal{J}^{\beta,k;\psi}_{a+} f(t) = \frac{\Gamma(\frac{\alpha}{k})\Gamma(\frac{\beta}{k})}{k^2 \Gamma_k(\alpha)\Gamma_k(\beta)} \mathcal{J}^{\frac{\alpha}{k};\psi}_{a+} \mathcal{J}^{\frac{\beta}{k};\psi}_{a+} f(t)$$

$$= \frac{\Gamma(\frac{\alpha}{k})\Gamma(\frac{\beta}{k})}{k^2 k^{\frac{\alpha}{k}-1} \Gamma(\frac{\alpha}{k}) k^{\frac{\beta}{k}-1} \Gamma(\frac{\beta}{k})} \mathcal{J}^{\frac{\alpha}{k};\psi}_{a+} \mathcal{J}^{\frac{\beta}{k};\psi}_{a+} f(t)$$

$$= \frac{1}{k^{\frac{\alpha+\beta}{k}}} \mathcal{J}^{\frac{\alpha+\beta}{k};\psi}_{a+} f(t)$$

$$= \mathcal{J}^{\alpha+\beta,k;\psi}_{a+} f(t),$$

where $\mathcal{J}^{\alpha;\psi}_{a+}$ is ψ-Riemann-Liouville fractional integral. We also have

$$\mathcal{J}^{\alpha,k;\psi}_{a+} \mathcal{J}^{\beta,k;\psi}_{a+} f(t) = \frac{\Gamma(\frac{\alpha}{k})\Gamma(\frac{\beta}{k})}{k^2 \Gamma_k(\alpha)\Gamma_k(\beta)} \mathcal{J}^{\frac{\alpha}{k};\psi}_{a+} \mathcal{J}^{\frac{\beta}{k};\psi}_{a+} f(t)$$

$$= \frac{\Gamma(\frac{\alpha}{k})\Gamma(\frac{\beta}{k})}{k^2 \Gamma_k(\alpha)\Gamma_k(\beta)} \mathcal{J}^{\frac{\beta}{k};\psi}_{a+} \mathcal{J}^{\frac{\alpha}{k};\psi}_{a+} f(t)$$

$$= \mathcal{J}^{\beta,k;\psi}_{a+} \mathcal{J}^{\alpha,k;\psi}_{a+} f(t).$$

$\square$

Lemma 2.19 ([33]) *Let $t > a$. Then, for $\alpha \geq 0$ and $\beta > 0 \, \rho > 0$, we have*

$$\left[{}^\rho \mathcal{J}^\alpha_{a+} \left(\frac{s^\rho - a^\rho}{\rho} \right)^{\beta-1} \right](t) = \frac{\Gamma(\beta)}{\Gamma(\alpha+\beta)} \left(\frac{t^\rho - a^\rho}{\rho} \right)^{\alpha+\beta-1},$$

$$\left[{}^\rho \mathcal{D}^\alpha_{a+} \left(\frac{s^\rho - a^\rho}{\rho} \right)^{\alpha-1} \right](t) = 0, \quad 0 < \alpha < 1.$$

Lemma 2.20 ([85, 148]) *Let $t > a$. Then, for $\alpha \geq 0$ and $\beta > 0$, we have*

$$\left[\mathcal{J}_{a+}^{\alpha;\psi}\left(\psi(\tau) - \psi(a)\right)^{\beta-1}\right](t) = \frac{\Gamma(\beta)}{\Gamma(\alpha+\beta)}\left(\psi(t) - \psi(a)\right)^{\alpha+\beta-1}.$$

Lemma 2.21 *Let $\alpha, \beta > 0$ and $k > 0$. Then, we have*

$$\mathcal{J}_{a+}^{\alpha,k;\psi}\left[\psi(t) - \psi(a)\right]^{\frac{\beta}{k}-1} = \frac{\Gamma_k(\beta)}{\Gamma_k(\alpha+\beta)}\left(\psi(t) - \psi(a)\right)^{\frac{\alpha+\beta}{k}-1}$$

and

$$\mathcal{J}_{b-}^{\alpha,k;\psi}\left[\psi(b) - \psi(t)\right]^{\frac{\beta}{k}-1} = \frac{\Gamma_k(\beta)}{\Gamma_k(\alpha+\beta)}\left(\psi(b) - \psi(t)\right)^{\frac{\alpha+\beta}{k}-1}.$$

Proof By Definition 2.7 and using the change of variable $\mu = \dfrac{\psi(s) - \psi(a)}{\psi(t) - \psi(a)}$, where $t > a$, we get

$$\mathcal{J}_{a+}^{\alpha,k;\psi}\left[\psi(t) - \psi(a)\right]^{\frac{\beta}{k}-1}$$

$$= \frac{1}{k\Gamma_k(\alpha)}\int_a^t \left(\psi(t) - \psi(s)\right)^{\frac{\alpha}{k}-1}\psi'(s)\left(\psi(s) - \psi(a)\right)^{\frac{\beta}{k}-1}ds$$

$$= \int_a^t \frac{\left(\psi(t) - \psi(a)\right)^{\frac{\alpha}{k}-1}}{k\Gamma_k(\alpha)}\left[1 - \frac{\psi(s) - \psi(a)}{\psi(t) - \psi(a)}\right]^{\frac{\alpha}{k}-1}\psi'(s)\left(\psi(s) - \psi(a)\right)^{\frac{\beta}{k}-1}ds$$

$$= \frac{\left[\psi(t) - \psi(a)\right]^{\frac{\alpha+\beta}{k}-1}}{k\Gamma_k(\alpha)}\int_0^1 \left[1 - \mu\right]^{\frac{\alpha}{k}-1}\mu^{\frac{\beta}{k}-1}d\mu.$$

Using the definition of k-beta function and the relation with gamma function, we have

$$\mathcal{J}_{a+}^{\alpha,k;\psi}\left[\psi(t) - \psi(a)\right]^{\frac{\beta}{k}-1} = \frac{\Gamma_k(\beta)}{\Gamma_k(\alpha+\beta)}\left(\psi(t) - \psi(a)\right)^{\frac{\alpha+\beta}{k}-1}.$$

□

Property 2.22 ([106]) The operator $^{\rho}\mathcal{D}_{a+}^{\alpha,\beta}$ can be written as

$$^{\rho}\mathcal{D}_{a+}^{\alpha,\beta} = {}^{\rho}\mathcal{J}_{a+}^{\beta(1-\alpha)}\delta_{\rho}\,{}^{\rho}\mathcal{J}_{a+}^{1-\gamma} = {}^{\rho}\mathcal{J}_{a+}^{\beta(1-\alpha)}\,{}^{\rho}\mathcal{D}_{a+}^{\gamma}, \quad \gamma = \alpha + \beta - \alpha\beta.$$

Lemma 2.23 ([85, 106]) *Let $\alpha > 0$, and $0 \leq \gamma < 1$. Then, $^{\rho}\mathcal{J}_{a+}^{\alpha}$ is bounded from $C_{\gamma,\rho}(J)$ into $C_{\gamma,\rho}(J)$. Since $^{\rho}\mathcal{D}_{a+}^{\alpha,\beta}u = {}^{\rho}\mathcal{J}_{a+}^{\beta(1-\alpha)}\,{}^{\rho}\mathcal{D}_{a+}^{\gamma}u$, it follows that*

$$C_{1-\gamma,\rho}^{\gamma}(J) \subset C_{1-\gamma,\rho}^{\alpha,\beta}(J) \subset C_{1-\gamma,\rho}(J).$$

Lemma 2.24 ([106]) *Let $0 < a < b < \infty$, $\alpha > 0$, $0 \leq \gamma < 1$, and $u \in C_{\gamma,\rho}(J)$. If $\alpha > 1 - \gamma$, then $^{\rho}\mathcal{J}_{a+}^{\alpha}u$ is continuous on J and*

$$\left({}^\rho \mathcal{J}^\alpha_{a+} u\right)(a) = \lim_{t\to a^+} \left({}^\rho \mathcal{J}^\alpha_{a+} u\right)(t) = 0.$$

Lemma 2.25 ([148]) *Let* $0 < a < b < \infty, \alpha > 0, 0 \le \gamma < 1, u \in C_{\gamma;\psi}(J)$. *If* $\alpha > 1 - \gamma$, *then* $\mathcal{J}^{\alpha;\psi}_{a+} u \in C(\bar{J}, \mathbb{R})$ *and*

$$\left(\mathcal{J}^{\alpha;\psi}_{a+} u\right)(a) = \lim_{t\to a^+} \left(\mathcal{J}^{\alpha;\psi}_{a+} u\right)(t) = 0.$$

Theorem 2.26 *Let* $0 < a < b < \infty, \alpha > 0, 0 \le \xi < 1, k > 0$, *and* $u \in C_{\xi;\psi}(J)$. *If* $\dfrac{\alpha}{k} > 1 - \xi$, *then*

$$\left(\mathcal{J}^{\alpha,k;\psi}_{a+} u\right)(a) = \lim_{t\to a^+} \left(\mathcal{J}^{\alpha,k;\psi}_{a+} u\right)(t) = 0.$$

Proof $u \in C_{\xi;\psi}(J)$ means that $(\psi(t) - \psi(a))^{1-\xi} u(t) \in C(J, \mathbb{R})$, then there exists a positive constant R such that

$$|(\psi(t) - \psi(a))^{1-\xi} u(t)| < R,$$

thus,

$$|u(t)| < R|(\psi(t) - \psi(a))^{\xi-1}|. \tag{2.1}$$

Now, we apply the operator $\mathcal{J}^{\alpha,k;\psi}_{a+}(\cdot)$ on both sides of Equation (2.1) and using Lemma 2.21, so that we have

$$\left|\left(\mathcal{J}^{\alpha,k;\psi}_{a+} u\right)(t)\right| < R \left|\mathcal{J}^{\alpha,k;\psi}_{a+} (\psi(t) - \psi(a))^{\xi-1}\right|$$

$$= \frac{R\Gamma_k(k\xi)}{\Gamma_k(\alpha + \xi)} (\psi(t) - \psi(a))^{\frac{\alpha}{k}+\xi-1}.$$

Then, we have the right-hand side $\to 0$ as $u \to a$, and

$$\lim_{t\to a^+} \left(\mathcal{J}^{\alpha,k;\psi}_{a+} u\right)(t) = \left(\mathcal{J}^{\alpha,k;\psi}_{a+} u\right)(a) = 0.$$

$\square$

Lemma 2.27 ([106]) *Let* $\alpha > 0, 0 \le \gamma < 1$, *and* $g \in C_{\gamma,\rho}(J)$. *Then,*

$$\left({}^\rho \mathcal{D}^\alpha_{a+} \, {}^\rho \mathcal{J}^\alpha_{a+} g\right)(t) = g(t), \quad \text{for all } t \in J.$$

Lemma 2.28 ([148]) *Let* $\alpha > 0, 0 \le \beta \le 1$, *and* $h \in C^1_{\gamma;\psi}(J)$. *Then,*

$$\left({}^H \mathcal{D}^{\alpha,\beta;\psi}_{a+} \, \mathcal{J}^{\alpha;\psi}_{a+} h\right)(t) = h(t), \quad \text{for all } t \in J.$$

Lemma 2.29 *Let $\alpha > 0, 0 \le \beta \le 1$, and $u \in C^1_{\xi;\psi}(J)$, where $k > 0$, then for $t \in J$, we have*

$$\left({}^H_k \mathcal{D}^{\alpha,\beta;\psi}_{a+} \mathcal{J}^{\alpha,k;\psi}_{a+} u \right)(t) = u(t).$$

Proof We have from Definition 2.12, Lemma 2.18, and $\xi = \frac{1}{k}(\beta(k - \alpha) + \alpha)$ that

$$\left({}^H_k \mathcal{D}^{\alpha,\beta;\psi}_{a+} \mathcal{J}^{\alpha,k;\psi}_{a+} u \right)(t) = \left(\mathcal{J}^{\beta(k-\alpha),k;\psi}_{a+} \delta^1_\psi \left(k \mathcal{J}^{(1-\beta)(k-\alpha),k;\psi}_{a+} \mathcal{J}^{\alpha,k;\psi}_{a+} u \right) \right)(t)$$

$$= \left(\mathcal{J}^{k\xi-\alpha,k;\psi}_{a+} \delta^1_\psi \left(k \mathcal{J}^{(1-\beta)(k-\alpha)+\alpha,k;\psi}_{a+} u \right) \right)(t)$$

$$= \left(\mathcal{J}^{k\xi-\alpha,k;\psi}_{a+} \delta^1_\psi \left(k \mathcal{J}^{k-k\xi+\alpha,k;\psi}_{a+} u \right) \right)(t),$$

then, we obtain

$$\left({}^H_k \mathcal{D}^{\alpha,\beta;\psi}_{a+} \mathcal{J}^{\alpha,k;\psi}_{a+} u \right)(t) = \frac{\int_a^t \frac{\psi'(s)}{(\psi(t) - \psi(s))^{1-\xi+\frac{\alpha}{k}}} \delta^1_\psi \left[\int_a^s \frac{\psi'(\tau)u(\tau)d\tau}{(\psi(s) - \psi(\tau))^{\xi-\frac{\alpha}{k}}} \right] ds}{k\Gamma_k(k\xi - \alpha)\Gamma_k(k(1 - \xi) + \alpha)}.$$

$$(2.2)$$

On other hand, by integrating by parts, we have

$$\int_a^s \frac{\psi'(\tau)u(\tau)d\tau}{(\psi(s) - \psi(\tau))^{\xi-\frac{\alpha}{k}}} = \frac{1}{1 - \xi + \frac{\alpha}{k}} \left[u(a) \left(\psi(s) - \psi(a) \right)^{1-\xi+\frac{\alpha}{k}} \right.$$

$$\left. + \int_a^s \frac{u'(\tau)d\tau}{(\psi(s) - \psi(\tau))^{\xi-1-\frac{\alpha}{k}}} \right],$$

then, by applying δ^1_ψ, we get

$$\delta^1_\psi \int_a^s \frac{\psi'(\tau)u(\tau)d\tau}{(\psi(s) - \psi(\tau))^{\xi-\frac{\alpha}{k}}} = u(a) \left(\psi(s) - \psi(a) \right)^{-\xi+\frac{\alpha}{k}} + \int_a^s \frac{u'(\tau)d\tau}{(\psi(s) - \psi(\tau))^{\xi-\frac{\alpha}{k}}}.$$

$$(2.3)$$

Now, replacing (2.3) into Equation (2.2), and by Dirichlet's formula and the properties of k-gamma function, we get

$$\left({}^H_k \mathcal{D}^{\alpha,\beta;\psi}_{a+} \mathcal{J}^{\alpha,k;\psi}_{a+} u \right)(t)$$

$$= \frac{1}{k\Gamma_k(k\xi - \alpha)\Gamma_k(k(1 - \xi) + \alpha)} \left[\int_a^t \frac{u(a)\psi'(s) \left(\psi(s) - \psi(a) \right)^{-\xi+\frac{\alpha}{k}} ds}{(\psi(t) - \psi(s))^{1-\xi+\frac{\alpha}{k}}} \right.$$

$$+ \int_a^t u'(t)dt \int_s^t \frac{\psi'(s)d\tau}{(\psi(t) - \psi(s))^{1-\xi+\frac{\alpha}{k}} (\psi(s) - \psi(\tau))^{\xi-\frac{\alpha}{k}}} \right].$$

Making the following change of variables $\mu = \dfrac{\psi(s) - \psi(a)}{\psi(t) - \psi(a)}$ in the integral from a to t and similarly changing the variable in the integral from s to t, then we have

$$\left({}^H_k \mathcal{D}^{\alpha,\beta;\psi}_{a+} \, \mathcal{J}^{\alpha,k;\psi}_{a+} u\right)(t)$$

$$= \frac{\displaystyle\int_a^t u(a)\psi'(s)\,(\psi(s) - \psi(a))^{-\xi+\frac{\alpha}{k}}\,(\psi(t) - \psi(s))^{\xi-\frac{\alpha}{k}-1}\,ds}{k\Gamma_k(k\xi - \alpha)\Gamma_k(k(1-\xi)+\alpha)}$$

$$+ \frac{\displaystyle\int_a^t u'(t)dt \int_s^t \psi'(s)\,(\psi(t) - \psi(s))^{\xi-\frac{\alpha}{k}-1}\,(\psi(s) - \psi(\tau))^{-\xi+\frac{\alpha}{k}}\,d\tau}{k\Gamma_k(k\xi - \alpha)\Gamma_k(k(1-\xi)+\alpha)}$$

$$= \frac{\left(u(a) + \displaystyle\int_a^t u'(t)dt\right)}{\Gamma_k(k\xi - \alpha)\Gamma_k(k(1-\xi)+\alpha)}\left[\frac{1}{k}\int_0^1 \mu^{-\xi+\frac{\alpha}{k}}\,(1-\mu)^{\xi-\frac{\alpha}{k}-1}\,d\mu\right]$$

$$= \frac{\left(u(a) + \displaystyle\int_a^t u'(t)dt\right)}{\Gamma_k(k\xi - \alpha)\Gamma_k(k(1-\xi)+\alpha)}\left[\frac{1}{k}\int_0^1 \mu^{(1-(\xi-\frac{\alpha}{k}))-1}\,(1-\mu)^{\xi-\frac{\alpha}{k}-1}\,d\mu\right],$$

then by the definition of k-beta function, we obtain

$$\left({}^H_k \mathcal{D}^{\alpha,\beta;\psi}_{a+} \, \mathcal{J}^{\alpha,k;\psi}_{a+} u\right)(t) = \frac{[\Gamma_k(k\xi - \alpha)\Gamma_k(k(1-\xi)+\alpha)]}{\Gamma_k(k\xi - \alpha)\Gamma_k(k(1-\xi)+\alpha)}\left(u(a) + \int_a^t u'(t)dt\right)$$

$$= u(a) + \int_a^t u'(t)dt$$

$$= u(t).$$

$\square$

Lemma 2.30 ([142, 148]) *Let $t > a$, $\alpha > 0, 0 \le \beta \le 1$. Then for $0 < \gamma < 1; \gamma = \alpha + \beta - \alpha\beta$, we have*

$$\left[\mathcal{D}^{\gamma;\psi}_{a+}\,(\psi(\tau) - \psi(a))^{\gamma-1}\right](t) = 0,$$

and

$$\left[{}^H\mathcal{D}^{\alpha,\beta;\psi}_{a+}\,(\psi(\tau) - \psi(a))^{\gamma-1}\right](t) = 0.$$

Lemma 2.31 *Let $t > a$, $\alpha > 0, 0 \le \beta \le 1, k > 0$. Then for $0 < \xi < 1; \xi = \frac{1}{k}(\beta(k - \alpha) + \alpha)$, we have*

$$\left[{}^H_k \mathcal{D}^{\alpha,\beta;\psi}_{a+}\,(\psi(s) - \psi(a))^{\xi-1}\right](t) = 0.$$

Proof From Definitions 2.7 and 2.12, we have

$$k\mathcal{J}_{a+}^{(1-\beta)(k-\alpha),k;\psi} \left(\psi(t) - \psi(a)\right)^{\xi-1} = \frac{k}{k\Gamma_k(kX)} \int_a^t \frac{\psi'(s)\left(\psi(s) - \psi(a)\right)^{\xi-1} ds}{\left(\psi(t) - \psi(s)\right)^{1-X}},$$

where $X = \dfrac{1}{k}(1-\beta)(k-\alpha)$. Now, we make the change of the variable by $\mu = \dfrac{\psi(s) - \psi(a)}{\psi(t) - \psi(a)}$ to obtain

$$k\mathcal{J}_{a+}^{(1-\beta)(k-\alpha),k;\psi} \left(\psi(t) - \psi(a)\right)^{\xi-1} = \frac{k\left[\psi(t) - \psi(a)\right]^{\xi-1+X}}{\Gamma_k(kX)}$$
$$\times \left[\frac{1}{k}\int_0^1 (1-\mu)^{X-1}\mu^{\xi-1}d\mu\right],$$

then, by the following definition of k-beta function

$$B_k(\alpha,\beta) = \frac{1}{k}\int_0^1 t^{\frac{\alpha}{k}-1}(1-t)^{\frac{\beta}{k}-1}dt = \frac{\Gamma_k(\alpha)\Gamma_k(\beta)}{\Gamma_k(\alpha+\beta)},$$

we have

$$k\mathcal{J}_{a+}^{(1-\beta)(k-\alpha),k;\psi} \left(\psi(t) - \psi(a)\right)^{\xi-1} = \frac{k\Gamma_k(k\xi)}{\Gamma_k(k(X+\xi))} = k\Gamma_k(k\xi),$$

$$k\mathcal{J}_{a+}^{(1-\beta)(k-\alpha),k;\psi} \left(\psi(t) - \psi(a)\right)^{\xi-1} = \frac{k\Gamma_k(k\xi)}{\Gamma_k(k(X+\xi))} = k\Gamma_k(k\xi),$$

then, we have

$$\delta_\psi^1 \left(k\mathcal{J}_{a+}^{(1-\beta)(k-\alpha),k;\psi} \left(\psi(t) - \psi(a)\right)^{\xi-1}\right) = 0.$$

$\square$

Lemma 2.32 ([106]) *Let* $0 < \alpha < 1, 0 \le \gamma < 1$. *If* $g \in C_{\gamma,\rho}(J)$ *and* $^\rho\mathcal{J}_{a+}^{1-\alpha}g \in C_{\gamma,\rho}^1(J)$, *then*

$$\left(^\rho\mathcal{J}_{a+}^\alpha\, ^\rho\mathcal{D}_{a+}^\alpha g\right)(t) = g(t) - \frac{\left(^\rho\mathcal{J}_{a+}^{1-\alpha}g\right)(a)}{\Gamma(\alpha)}\left(\frac{t^\rho - a^\rho}{\rho}\right)^{\alpha-1}, \quad \text{for all } t \in J.$$

Lemma 2.33 ([106]) *Let* $0 < \alpha < 1, 0 \le \beta \le 1$ *and* $\gamma = \alpha + \beta - \alpha\beta$. *If* $u \in C_{\gamma,\rho}^\gamma(J)$, *then*

$$^\rho\mathcal{J}_{a+}^\gamma\, ^\rho\mathcal{D}_{a+}^\gamma u = \, ^\rho\mathcal{J}_{a+}^\alpha\, ^\rho\mathcal{D}_{a+}^{\alpha,\beta} u,$$

and

$$\,^{\rho}\mathcal{D}_{a+}^{\gamma}\,\,^{\rho}\mathcal{J}_{a+}^{\alpha}u = \,^{\rho}\mathcal{D}_{a+}^{\beta(1-\alpha)}u.$$

Lemma 2.34 ([142, 148]) *Let* $\alpha > 0, 0 \le \beta \le 1$, *and* $h \in C^1_{\gamma;\psi}(J)$. *Then,*

$$\left(\mathcal{J}_{a+}^{\alpha;\psi} \,{}^{H}\mathcal{D}_{a+}^{\alpha,\beta;\psi}h\right)(t) = h(t) - \frac{\left(\mathcal{J}_{a+}^{1-\gamma;\psi}h\right)(a)}{\Gamma(\gamma)}\,(\psi(t) - \psi(a))^{\gamma-1}, \quad \text{for all } t \in J.$$

Theorem 2.35 *If* $f \in C^n_{\xi;\psi}[a,b], n - 1 < \alpha < n,\ 0 \le \beta \le 1$, *where* $n \in \mathbb{N}$ *and* $k > 0$, *then*

$$\left(\mathcal{J}_{a+}^{\alpha,k;\psi}\,{}^{H}_{k}\mathcal{D}_{a+}^{\alpha,\beta;\psi}f\right)(t) = \sum_{i=1}^{n} \frac{-(\psi(t)-\psi(a))^{\xi-i}}{k^{i-n}\Gamma_k(k(\xi-i+1))}\left\{\delta_\psi^{n-i}\left(\mathcal{J}_{a+}^{k(n-\xi),k;\psi}f(a)\right)\right\}$$
$$+ f(t),$$

where

$$\xi = \frac{1}{k}\left(\beta(kn - \alpha) + \alpha\right).$$

In particular, if $n = 1$, *we have*

$$\left(\mathcal{J}_{a+}^{\alpha,k;\psi}\,{}^{H}_{k}\mathcal{D}_{a+}^{\alpha,\beta;\psi}f\right)(t) = f(t) - \frac{(\psi(t)-\psi(a))^{\xi-1}}{\Gamma_k(\beta(k-\alpha)+\alpha)}\mathcal{J}_{a+}^{(1-\beta)(k-\alpha),k;\psi}f(a).$$

Proof From Definition 2.12 and Lemma 2.18, we have

$$\left(\mathcal{J}_{a+}^{\alpha,k;\psi}\,{}^{H}_{k}\mathcal{D}_{a+}^{\alpha,\beta;\psi}f\right)(t) = \left(\mathcal{J}_{a+}^{\alpha,k;\psi}\,\mathcal{J}_{a+}^{\beta(kn-\alpha),k;\psi}\delta_\psi^n\left(k^n\mathcal{J}_{a+}^{(1-\beta)(kn-\alpha),k;\psi}f\right)\right)(t)$$
$$= \left(\mathcal{J}_{a+}^{\beta(kn-\alpha)+\alpha,k;\psi}\delta_\psi^n\left(k^n\mathcal{J}_{a+}^{(1-\beta)(kn-\alpha),k;\psi}f\right)\right)(t)$$
$$= \frac{1}{k\Gamma_k(k\xi)}\int_a^t \frac{\psi'(s)\left\{\delta_\psi^n\left(k^n\mathcal{J}_{a+}^{(1-\beta)(kn-\alpha),k;\psi}f(s)\right)\right\}}{(\psi(t)-\psi(s))^{1-\xi}}ds.$$

Integrating by parts, we obtain

$$\left(\mathcal{J}_{a+}^{\alpha,k;\psi}\,{}^{H}_{k}\mathcal{D}_{a+}^{\alpha,\beta;\psi}f\right)(t) = \frac{-(\psi(t)-\psi(a))^{\xi-1}}{k\Gamma_k(k\xi)}\left\{\delta_\psi^{n-1}\left(k^n\mathcal{J}_{a+}^{(1-\beta)(kn-\alpha),k;\psi}f(a)\right)\right\}$$
$$+ \int_a^t \frac{(\xi-1)\psi'(s)\left\{\delta_\psi^{n-1}\left(k^n\mathcal{J}_{a+}^{(1-\beta)(kn-\alpha),k;\psi}f(s)\right)\right\}}{k\Gamma_k(k\xi)(\psi(t)-\psi(s))^{2-\xi}}ds.$$

Using the property of the functions gamma and k-gamma, we get

$$\left(\mathcal{J}_{a+}^{\alpha,k;\psi}\ {}^{H}_{k}\mathcal{D}_{a+}^{\alpha,\beta;\psi}f\right)(t) = \frac{-(\psi(t)-\psi(a))^{\xi-1}}{k^{\xi}\Gamma(\xi)}\left\{\delta_{\psi}^{n-1}\left(k^{n}\mathcal{J}_{a+}^{(1-\beta)(kn-\alpha),k;\psi}f(a)\right)\right\}$$

$$+\int_{a}^{t}\frac{\psi'(s)\left\{\delta_{\psi}^{n-1}\left(k^{n}\mathcal{J}_{a+}^{(1-\beta)(kn-\alpha),k;\psi}f(s)\right)\right\}}{k^{\xi}\Gamma(\xi-1)\left(\psi(t)-\psi(s)\right)^{2-\xi}}ds.$$

So, by integrating by parts n times, we obtain

$$\left(\mathcal{J}_{a+}^{\alpha,k;\psi}\ {}^{H}_{k}\mathcal{D}_{a+}^{\alpha,\beta;\psi}f\right)(t)$$

$$= -\sum_{i=1}^{n}\frac{(\psi(t)-\psi(a))^{\xi-i}}{k^{\xi}\Gamma(\xi-i+1)}\left\{\delta_{\psi}^{n-i}\left(k^{n}\mathcal{J}_{a+}^{(1-\beta)(kn-\alpha),k;\psi}f(a)\right)\right\}$$

$$+\frac{1}{k^{\xi-n}\Gamma(\xi-n)}\int_{a}^{t}\frac{\psi'(s)}{(\psi(t)-\psi(s))^{n+1-\xi}}\left(\mathcal{J}_{a+}^{(1-\beta)(kn-\alpha),k;\psi}f(s)\right)ds,$$

$$= -\sum_{i=1}^{n}\frac{(\psi(t)-\psi(a))^{\xi-i}}{k^{i}\Gamma_{k}(k(\xi-i+1))}\left\{\delta_{\psi}^{n-i}\left(k^{n}\mathcal{J}_{a+}^{(1-\beta)(kn-\alpha),k;\psi}f(a)\right)\right\}$$

$$+\frac{1}{k\Gamma_{k}(k(\xi-n))}\int_{a}^{t}\frac{\psi'(s)}{(\psi(t)-\psi(s))^{n+1-\xi}}\left(\mathcal{J}_{a+}^{(1-\beta)(kn-\alpha),k;\psi}f(s)\right)ds,$$

$$= -\sum_{i=1}^{n}\frac{(\psi(t)-\psi(a))^{\xi-i}}{k^{i-n}\Gamma_{k}(k(\xi-i+1))}\left\{\delta_{\psi}^{n-i}\left(\mathcal{J}_{a+}^{(1-\beta)(kn-\alpha),k;\psi}f(a)\right)\right\}$$

$$+\mathcal{J}_{a+}^{k(\xi-n),k;\psi}\mathcal{J}_{a+}^{(1-\beta)(kn-\alpha),k;\psi}f(t),$$

then by using Lemma 2.18, we get

$$\left(\mathcal{J}_{a+}^{\alpha,k;\psi}\ {}^{H}_{k}\mathcal{D}_{a+}^{\alpha,\beta;\psi}f\right)(t) = \sum_{i=1}^{n}\frac{-(\psi(t)-\psi(a))^{\xi-i}\delta_{\psi}^{n-i}\left(\mathcal{J}_{a+}^{(1-\beta)(kn-\alpha),k;\psi}f(a)\right)}{k^{i-n}\Gamma_{k}(k(\xi-i+1))}$$

$$+ f(t). \qquad\qquad\qquad \square$$

Property 2.36 ([148]) The operator ${}^{H}\mathcal{D}_{a+}^{\alpha,\beta;\psi}$ can be written as

$$^{H}\mathcal{D}_{a+}^{\alpha,\beta;\psi} = \mathcal{J}_{a+}^{\beta(1-\alpha);\psi}\,\mathcal{D}_{a+}^{\gamma;\psi},\quad \gamma = \alpha + \beta - \alpha\beta.$$

Lemma 2.37 ([85, 148]) *Let $\alpha > 0$, $\beta > 0, 0 < a < b < \infty$. Then, for $h \in X_{c}^{p}(a, b)$ the semigroup property is valid, i.e.,*

$$\left(\mathcal{J}_{a+}^{\alpha;\psi}\,\mathcal{J}_{a+}^{\beta;\psi}h\right)(t) = \left(\mathcal{J}_{a+}^{\alpha+\beta;\psi}h\right)(t).$$

Lemma 2.38 ([106]) *Let* f *be a function such that* $f \in C_{\gamma,\rho}(J)$. *Then* $u \in C_{\gamma,\rho}^{\gamma}(J)$ *is a solution of the differential equation:*

$$\left({}^{\rho}\mathcal{D}_{a+}^{\alpha,\beta} u \right)(t) = f(t), \text{ for each }, \ t \in J, \ 0 < \alpha < 1, \ 0 \le \beta \le 1, \rho > 0$$

if and only if u *satisfies the following Volterra integral equation:*

$$u(t) = \frac{\left({}^{\rho}\mathcal{J}_{a+}^{1-\gamma} u \right)(a^{+})}{\Gamma(\gamma)} \left(\frac{t^{\rho} - a^{\rho}}{\rho} \right)^{\gamma-1} + \frac{1}{\Gamma(\alpha)} \int_{a}^{t} \left(\frac{t^{\rho} - s^{\rho}}{\rho} \right)^{\alpha-1} s^{\rho-1} f(s)ds,$$

where $\gamma = \alpha + \beta - \alpha\beta$.

Lemma 2.39 *Let* $\alpha, \beta > 0$ *and* $k > 0$. *Then, we have*

$$\mathcal{J}_{a+}^{\alpha,k;\psi} \mathbb{E}_{k}^{\alpha} \left((\psi(t) - \psi(a))^{\frac{\alpha}{k}} \right) = \mathbb{E}_{k}^{\alpha} \left((\psi(t) - \psi(a))^{\frac{\alpha}{k}} \right) - 1.$$

Proof We have

$$\mathcal{J}_{a+}^{\alpha,k;\psi} \mathbb{E}_{k}^{\alpha} \left((\psi(t) - \psi(a))^{\frac{\alpha}{k}} \right)$$

$$= \frac{1}{k\Gamma_{k}(\alpha)} \int_{a}^{t} \frac{\psi'(s)\mathbb{E}_{k}^{\alpha}\left((\psi(s) - \psi(a))^{\frac{\alpha}{k}} \right) ds}{(\psi(t) - \psi(s))^{1-\frac{\alpha}{k}}}$$

$$= \frac{1}{k\Gamma_{k}(\alpha)} \int_{a}^{t} \frac{\psi'(s)}{(\psi(t) - \psi(s))^{1-\frac{\alpha}{k}}} \sum_{i=0}^{\infty} \frac{(\psi(s) - \psi(a))^{\frac{\alpha i}{k}}}{\Gamma_{k}(\alpha i + k)} ds.$$

With a change of variables $\mu = \psi(s) - \psi(a)$, we get

$$\mathcal{J}_{a+}^{\alpha,k;\psi} \mathbb{E}_{k}^{\alpha} \left((\psi(t) - \psi(a))^{\frac{\alpha}{k}} \right)$$

$$= \frac{1}{k\Gamma_{k}(\alpha)} \sum_{i=0}^{\infty} \frac{1}{\Gamma_{k}(\alpha i + k)} \int_{0}^{\psi(t)-\psi(a)} \frac{\mu^{\frac{\alpha i}{k}}}{(\psi(t) - \psi(a) - \mu)^{1-\frac{\alpha}{k}}} d\mu$$

$$= \frac{1}{k\Gamma_{k}(\alpha)} \sum_{i=0}^{\infty} \frac{(\psi(t) - \psi(a))^{\frac{\alpha}{k}-1}}{\Gamma_{k}(\alpha i + k)} \int_{0}^{\psi(t)-\psi(a)} \mu^{\frac{\alpha i}{k}} \left(1 - \frac{\mu}{\psi(t) - \psi(a)} \right)^{\frac{\alpha}{k}-1} d\mu.$$

Making the change of variables $\varsigma = \dfrac{\mu}{\psi(t) - \psi(a)}$ and using the definition of k-beta function, we have

$$\mathcal{J}_{a+}^{\alpha,k;\psi} \mathbb{E}_k^\alpha \left((\psi(t) - \psi(a))^{\frac{\alpha}{k}} \right)$$

$$= \frac{1}{k\Gamma_k(\alpha)} \sum_{i=0}^{\infty} \frac{(\psi(t) - \psi(a))^{\frac{\alpha}{k}(i+1)}}{\Gamma_k(\alpha i + k)} \int_0^1 \varsigma^{\frac{\alpha i}{k}} (1 - \varsigma)^{\frac{\alpha}{k}-1} d\varsigma$$

$$= \frac{1}{\Gamma_k(\alpha)} \sum_{i=0}^{\infty} \frac{(\psi(t) - \psi(a))^{\frac{\alpha}{k}(i+1)} \Gamma_k(\alpha i + k) \Gamma_k(\alpha)}{\Gamma_k(\alpha i + k) \Gamma_k(\alpha(i+1) + k)}$$

$$= \sum_{i=0}^{\infty} \frac{(\psi(t) - \psi(a))^{\frac{\alpha}{k}(i+1)}}{\Gamma_k(\alpha(i+1) + k)}$$

$$= \sum_{j=0}^{\infty} \frac{(\psi(t) - \psi(a))^{\frac{\alpha j}{k}}}{\Gamma_k(\alpha j + k)} - 1 = \mathbb{E}_k^\alpha \left((\psi(t) - \psi(a))^{\frac{\alpha}{k}} \right) - 1.$$

$\square$

2.4 Grönwall's Lemma

The Grönwall's inequality is fundamental in the study of qualitative theory of integral and differential equations, as well as in the solution of Cauchy-type problems of nonlinear differential equations.

Lemma 2.40 (**Grönwall's** lemma [32]) *Let u and w be two integrable functions and v be a continuous function, with domain $\bar{J}$. Assume that*

- *u and w are nonnegative;*
- *v is nonnegative and nondecreasing.*

If

$$u(t) \le w(t) + v(t) \int_a^t s^{\rho-1} \left(\frac{t^\rho - s^\rho}{\rho} \right)^{\alpha-1} u(s) ds, \ t \in \bar{J},$$

then

$$u(t) \le w(t) + \int_a^t \sum_{\tau=1}^{\infty} \frac{(v(t)\Gamma(\alpha))^\tau}{\Gamma(\tau\alpha)} s^{\rho-1} \left(\frac{t^\rho - s^\rho}{\rho} \right)^{\tau\alpha-1} w(s) ds, \ t \in \bar{J}.$$

In addition, if w is nondecreasing, then

$$u(t) \le w(t) \mathbb{E}_\alpha \left[v(t)\Gamma(\alpha) \left(\frac{t^\rho - a^\rho}{\rho} \right)^\alpha \right], \ t \in \bar{J}.$$

In 2019, Sousa *et al.* managed to give a Grönwall's inequality using the ψ-Hilfer fractional integral.

Lemma 2.41 ([145]) *Let u, v be two integrable functions and g continuous, with domain $[a, b]$. Let $\psi \in C^1[a, b]$ be an increasing function such that $\psi'(t) \neq 0$, for all $t \in [a, b]$. Assume that*

- *u and v are nonnegative;*
- *g in nonnegative and nondecreasing.*

If

$$u(t) \leq v(t) + g(t) \int_a^t \psi'(\tau)(\psi(t) - \psi(\tau))^{\alpha-1} u(\tau) d\tau,$$

then

$$u(t) \leq v(t) + \int_a^t \sum_{k=1}^\infty \frac{[g(t)\Gamma(\alpha)]^k}{\Gamma(\alpha k)} \psi'(\tau)[\psi(t) - \psi(\tau)]^{\alpha k-1} v(\tau) d\tau,$$

for all $t \in [a, b]$.

Now, we give generalized Grönwall's inequality taking into account the properties of the functions k-gamma, k-beta, and k-Mittag-Leffler.

Theorem 2.42 *Let u, v be two integrable functions and g continuous, with domain $\bar{J}$. Let $\psi \in C^1(J)$ be an increasing function such that $\psi'(t) \neq 0$, $t \in \bar{J}$, and $\alpha > 0$ with $k > 0$. Assume that*

1. *u and v are nonnegative;*
2. *w is nonnegative and nondecreasing.*

If

$$u(t) \leq v(t) + \frac{w(t)}{k} \int_a^t \psi'(s) [\psi(t) - \psi(s)]^{\frac{\alpha}{k}-1} u(s) ds,$$

then

$$u(t) \leq v(t) + \int_a^t \sum_{i=1}^\infty \frac{[w(t) \Gamma_k(\alpha)]^i}{k\Gamma_k(\alpha i)} \psi'(s) [\psi(t) - \psi(s)]^{\frac{i\alpha}{k}-1} v(s) ds, \qquad (2.4)$$

for all $t \in \bar{J}$. And if v is a nondecreasing function on $\bar{J}$, then we have

$$u(t) \leq v(t) \mathbb{E}_k^{\alpha,k} \left(w(t) \Gamma_k(\alpha) (\psi(t) - \psi(a))^{\frac{\alpha}{k}} \right).$$

Proof Let

$$\Upsilon v(t)(t) = \frac{w(t)}{k} \int_a^t \psi'(\tau) [\psi(t) - \psi(\tau)]^{\frac{\alpha}{k}-1} v(\tau) d\tau, \qquad (2.5)$$

for all $t \in \bar{J}$, for locally integral function v, we have

$$u(t) \le v(t) + \Upsilon u(t).$$

Iterating, for $n \in \mathbb{N}$, we can write

$$u(t) \le \sum_{i=0}^{n-1} \Upsilon^i v(t) + \Upsilon^n u(t).$$

Then, by mathematical induction, and if v is a nonnegative function, then

$$\Upsilon^n u(t) \le \int_a^t \frac{[w(t) \Gamma_k(\alpha)]^n}{k \Gamma_k(n\alpha)} \psi'(\tau) [\psi(t) - \psi(\tau)]^{\frac{n\alpha}{k} - 1} u(\tau) d\tau. \tag{2.6}$$

We know that relation Eq. (2.6) is true for $n = 1$. Suppose that the formula is true for some $n = i \in \mathbb{N}$, then the induction hypothesis implies

$$\Upsilon^{i+1} u(t) = \Upsilon \left(\Upsilon^i u(t) \right)$$

$$\le \Upsilon \left(\int_a^t \frac{[w(t) \Gamma_k(\alpha)]^i}{k \Gamma_k(i\alpha)} \psi'(\tau) [\psi(t) - \psi(\tau)]^{\frac{i\alpha}{k} - 1} v(\tau) d\tau \right)$$

$$= \frac{w(t)}{k} \int_a^t \frac{\psi'(\tau)}{[\psi(t) - \psi(\tau)]^{1 - \frac{\alpha}{k}}} \left(\int_a^\tau \frac{[w(\tau) \Gamma_k(\alpha)]^i \psi'(s) v(s) ds}{k \Gamma_k(i\alpha) [\psi(\tau) - \psi(s)]^{1 - \frac{i\alpha}{k}}} \right) d\tau.$$

Since w is a nondecreasing function, that is $w(\tau) \le w(t)$, for all $\tau \le t$, then we obtain

$$\Upsilon^{i+1} u(t) \le \frac{\Gamma_k(\alpha)^i}{k^2 \Gamma_k(\alpha i)} [w(t)]^{i+1}$$

$$\times \int_a^t \int_a^\tau \psi'(\tau) [\psi(t) - \psi(\tau)]^{\frac{\alpha}{k} - 1} \psi'(s) [\psi(\tau) - \psi(s)]^{\frac{i\alpha}{k} - 1} u(s) ds d\tau.$$

$$\tag{2.7}$$

From Eq. (2.7) and by Dirichlet's formula, we can have

$$\Upsilon^{i+1} u(t) \le \frac{\Gamma_k(\alpha)^i}{k^2 \Gamma_k(\alpha i)} [w(t)]^{i+1}$$

$$\times \int_a^t \psi'(\tau) u(\tau) \int_\tau^t \psi'(s) [\psi(t) - \psi(s)]^{\frac{\alpha}{k} - 1} [\psi(s) - \psi(\tau)]^{\frac{i\alpha}{k} - 1} ds d\tau.$$

$$\tag{2.8}$$

On other hand, we have

$$\int_\tau^t \psi'(s) \, [\psi(t) - \psi(s)]^{\frac{\alpha}{k}-1} \, [\psi(s) - \psi(\tau)]^{\frac{i\alpha}{k}-1} \, ds$$

$$= \int_\tau^t \psi'(s) \, [\psi(t) - \psi(\tau)]^{\frac{\alpha}{k}-1} \left[1 - \frac{\psi(s) - \psi(\tau)}{\psi(t) - \psi(\tau)}\right]^{\frac{\alpha}{k}-1} [\psi(s) - \psi(\tau)]^{\frac{i\alpha}{k}-1} \, ds.$$

With a change of variables $\mu = \dfrac{\psi(s) - \psi(\tau)}{\psi(t) - \psi(\tau)}$ and using the definition of k-beta function

and the relation with gamma function $B_k(\alpha, \beta) = \dfrac{\Gamma_k(\alpha) \, \Gamma_k(\beta)}{\Gamma_k(\alpha + \beta)}$, we have

$$\int_\tau^t \psi'(s) \, [\psi(t) - \psi(s)]^{\frac{\alpha}{k}-1} \, [\psi(s) - \psi(\tau)]^{\frac{i\alpha}{k}-1} \, ds$$

$$= [\psi(t) - \psi(\tau)]^{\frac{i\alpha+\alpha}{k}-1} \int_0^1 [1 - \mu]^{\frac{\alpha}{k}-1} \, \mu^{\frac{i\alpha}{k}-1} d\mu$$

$$= k \, [\psi(t) - \psi(\tau)]^{\frac{i\alpha+\alpha}{k}-1} \frac{\Gamma_k(\alpha) \, \Gamma_k(i\alpha)}{\Gamma_k(\alpha + i\alpha)}. \tag{2.9}$$

By replacing Eq. (2.9) in Eq. (2.8), we get

$$\Upsilon^{i+1} u(t) \le \int_a^t \frac{[w(t) \, \Gamma_k(\alpha)]^{i+1}}{k\Gamma_k(\alpha(i+1))} \psi'(\tau) \, u(\tau) \, [\psi(t) - \psi(\tau)]^{\frac{\alpha(i+1)}{k}-1} \, d\tau.$$

Let us now prove that $\Upsilon^n u(t) \to 0$ as $n \to \infty$. Since w is a continuous function on $\bar{J}$, there exists a constant $M > 0$ such that $w(t) \le M$ for all $t \in \bar{J}$. Then, we obtain

$$\Upsilon^n u(t) \le \int_a^t \frac{[M\Gamma_k(\alpha)]^n}{k\Gamma_k(\alpha n)} \psi'(\tau) \, u(\tau) \, [\psi(t) - \psi(\tau)]^{\frac{\alpha n}{k}-1} \, d\tau.$$

Consider the series

$$\sum_{n=1}^\infty \frac{[M\Gamma_k(\alpha)]^n}{\Gamma_k(\alpha n)}.$$

Using the property of the generalized k-gamma, we have

$$\sum_{n=1}^\infty \frac{[M\Gamma_k(\alpha)]^n}{\Gamma_k(\alpha n)} = \sum_{n=1}^\infty \frac{\left[Mk^{\frac{\alpha}{k}-1}\Gamma\left(\frac{\alpha}{k}\right)\right]^n}{k^{\frac{\alpha n}{k}-1}\Gamma\left(\frac{\alpha n}{k}\right)} = \sum_{n=1}^\infty \frac{k\left[Mk^{-1}\Gamma\left(\frac{\alpha}{k}\right)\right]^n}{\Gamma\left(\frac{\alpha n}{k}\right)}.$$

By using the Stirling approximation and the root test, we can show that the series converges. Therefore, we conclude that

$$u(t) \leq \sum_{i=0}^{\infty} \Upsilon^i v(t)$$

$$\leq v(t) + \int_a^t \sum_{i=1}^{\infty} \frac{[w(t)\, \Gamma_k(\alpha)]^i}{k \Gamma_k(\alpha i)} \psi'(\tau) [\psi(t) - \psi(\tau)]^{\frac{\alpha i}{k} - 1} v(\tau)\, d\tau.$$

Now, since v is nondecreasing, so, for all $\tau \in [a, t]$, we have $v(\tau) \leq v(t)$ and we can write

$$u(t) \leq v(t) + \int_a^t \sum_{i=1}^{\infty} \frac{[w(t)\, \Gamma_k(\alpha)]^i}{k \Gamma_k(\alpha i)} \psi'(\tau) [\psi(t) - \psi(\tau)]^{\frac{\alpha i}{k} - 1} v(\tau)\, d\tau$$

$$\leq v(t) \left[1 + \int_a^t \sum_{i=1}^{\infty} \frac{[w(t)\, \Gamma_k(\alpha)]^i}{k \Gamma_k(\alpha i)} \psi'(\tau) [\psi(t) - \psi(\tau)]^{\frac{\alpha i}{k} - 1} d\tau \right]$$

$$= v(t) \left[1 + \sum_{i=1}^{\infty} \frac{[w(t)\, \Gamma_k(\alpha)]^i}{\alpha i\, \Gamma_k(\alpha i)} [\psi(t) - \psi(a)]^{\frac{\alpha i}{k}} \right],$$

and by using the properties of k-gamma function and the definition of k-Mittag-Leffler function, we have

$$u(t) \leq v(t) \left[1 + \sum_{i=1}^{\infty} \frac{\left[w(t)\, \Gamma_k(\alpha)\, (\psi(t) - \psi(a))^{\frac{\alpha}{k}} \right]^i}{\Gamma_k(\alpha i + k)} \right]$$

$$= v(t)\, \mathbb{E}_k^{\alpha, k} \left(w(t)\, \Gamma_k(\alpha)\, (\psi(t) - \psi(a))^{\frac{\alpha}{k}} \right).$$

$\square$

2.5   Kuratowski Measure of Noncompactness

As mentioned in the introduction part, the measure of noncompactness is one of the fundamental tools in the theory of nonlinear analysis. In this section, we recall some fundamental facts of the notion of measure of noncompactness. Particularly, we employ the Kuratowski measure of noncompactness in our studies throughout this book.

Let Ω_X be the class of all bounded subsets of a metric space X.

Definition 2.43 ([45]) A function $\mu : \Omega_X \to [0, \infty)$ is said to be a measure of noncompactness on X if the following conditions are verified for all $B, B_1, B_2 \in \Omega_X$.

(a) Regularity, i.e., $\mu(B) = 0$ if and only if B is precompact,
(b) invariance under closure, i.e., $\mu(B) = \mu(\overline{B})$,
(c) semi-additivity, i.e., $\mu(B_1 \cup B_2) = \max\{\mu(B_1), \mu(B_2)\}$.

Definition 2.44 ([45]) Let X be a Banach space. The Kuratowski measure of noncompact-ness is the map $\mu : \Omega_X \longrightarrow [0, \infty)$ defined by

$$\mu(M) = inf\{\epsilon > 0 : M \subset \bigcup_{j=1}^{m} M_j, diam(M_j) \leq \epsilon\},$$

where $M \in \Omega_X$.
The map μ satisfies the following properties :

- $\mu(M) = 0 \Leftrightarrow \overline{M}$ is compact (M is relatively compact).
- $\mu(M) = \mu(\overline{M})$.
- $M_1 \subset M_2 \Rightarrow \mu(M_1) \leq \mu(M_2)$.
- $\mu(M_1 + M_2) \leq \mu(B_1) + \mu(B_2)$.
- $\mu(cM) = |c|\mu(M), c \in \mathbb{R}$.
- $\mu(convM) = \mu(M)$.

2.6   Fixed Point Theorems

In this section, we will be going through all the fixed point theorems used in the different studies throughout the monograph. Fixed point theory is one of the most intensively studied research topics of the last decades. The roots of the fixed point concept date back to the middle of the eighteenth century. Although fixed point theory appears to be an independent research topic today, the notion of the fixed point appeared in the papers that dealt with the solution of certain differential equations; see, e.g., Liouville [93], Picard [110], and Poincaré [112]. One of the first independent fixed point results was obtained by Banach [44] by abstracting the successive approximation method of Picard.

Theorem 2.45 (**Banach's** fixed point theorem [72]) *Let D be a nonempty closed subset of a Banach space E, then any contraction mapping N of D into itself has a unique fixed point.*

The most important feature that distinguishes **Banach's** fixed point theorem from other fixed point theorems is that it guarantees not only the existence but also the uniqueness of the fixed point. More importantly, it not only tells the existence and uniqueness of a fixed point but also tells you how to get the fixed point.

In what follows, we list some other fixed point theorems that have turned out to be the instruments for the differential equations solutions.

Theorem 2.46 (*Schauder's* fixed point theorem [72]) *Let X be a Banach space, D be a bounded closed convex subset of X, and $T : D \to D$ be a compact and continuous map. Then T has at least one fixed point in D.*

Theorem 2.47 (*Schaefer's* fixed point theorem [72]) *Let X be a Banach space and $N : X \to X$ be a completely continuous operator. If the set*

$$D = \{u \in X : u = \lambda Nu, \text{ for some } \lambda \in (0, 1)\}$$

is bounded, then N has a fixed point.

Theorem 2.48 (*Darbo's* fixed point Theorem [71]) *Let D be a nonempty, closed, bounded, and convex subset of a Banach space X, and let T be a continuous mapping of D into itself such that for any nonempty subset C of D,*

$$\mu(T(C)) \leq k\mu(C), \tag{2.10}$$

where $0 \leq k < 1$, and μ is the Kuratowski measure of noncompactness. Then T has a fixed point in D.

Theorem 2.49 (*Mönch's* fixed point Theorem [101]) *Let D be closed, bounded, and convex subset of a Banach space X such that $0 \in D$, and let T be a continuous mapping of D into itself. If the implication*

$$V = \overline{conv}T(V), \quad or \quad V = T(V) \cup \{0\} \Rightarrow \mu(V) = 0, \tag{2.11}$$

holds for every subset V of D, then T has a fixed point.

Theorem 2.50 (*Krasnoselskii's* fixed point theorem [72]) *Let D be a closed, convex, and nonempty subset of a Banach space X, and A, B the operators such that*

(1) $Au + Bv \in D$ for all $u, v \in D$;
(2) A is compact and continuous;
(3) B is a contraction mapping.

   Then there exists $w \in D$ such that $w = Aw + Bw$.

Theorem 2.51 *Let Ω be a closed, convex, bounded, and nonempty subset of a Banach algebra $(X, \| \cdot \|)$, and let $T_1 : X \to X$ and $T_2 : \Omega \to X$ be two operators such that*

(1) T_1 is Lipschitzian with Lipschitz constant λ,
(2) T_2 is completely continuous,
(3) $v = T_1 y T_2 w \Rightarrow v \in \Omega$ for all $w \in \Omega$,
(4) $\lambda \mathcal{M} < 1$, where $\mathcal{M} = \|B(\Omega)\| = \sup\{\|B(w)\| : w \in \Omega\}$.

Then the operator equation $T_1 y T_2 y = v$ has a solution in Ω.

Theorem 2.52 *Let B be a closed, convex, bounded and nonempty subset of a Banach algebra $(X, \|\cdot\|)$, and let $\mathcal{P}, \mathcal{R} : X \to X$ and $\mathcal{Q} : B \to X$ be three operators such that*

(1) $\mathcal{P}$ and $\mathcal{R}$ are Lipschitzian with Lipschitz constants η_1 and η_2, respectively,
(2) $\mathcal{Q}$ is compact and continuous,
(3) $u = \mathcal{P}u\mathcal{Q}v + \mathcal{R}u \Rightarrow u \in B$ for all $v \in B$
(4) $\eta_1 \beta + \eta_2 < 1$, where $\beta = \|\mathcal{Q}(B)\| = \sup\{\|\mathcal{Q}(v)\| : v \in B\}$.

Then the operator equation $\mathcal{P}u\mathcal{Q}u + \mathcal{R}u = u$ has a solution in B.

Implicit Fractional Differential Equations

3

3.1 Introduction and Motivations

This chapter deals with some existence and Ulam stability results for a class of initial and boundary value problems for differential equations with generalized Hilfer-type fractional derivative in Banach spaces. The results are based on suitable fixed point theorems associated with the technique of measure of noncompactness. At the end of each section, examples are included to show the applicability of our results. The results obtained in this chapter are studied and presented as a consequence of the following:

- The monographs of Abbas et al. [7, 14] and Benchohra et al. [50] and the papers of Ahmad et al. [25], Benchohra et al. [48, 49], and Zhou et al. [162], which are focused on linear and nonlinear initial and boundary value problems for fractional differential equations involving different kinds of fractional derivatives.
- The monographs of Abbas et al. [7, 13], Kilbas et al. [85], and Zhou et al. [162], and the papers of Abbas et al. [10] and Benchohra et al. [51, 52]; in it, considerable attention has been given to the study of the Ulam-Hyers and Ulam-Hyers-Rassias stability of various classes of functional equations.
- The paper of Diaz et al. [66], where the authors presented the k-gamma and k-beta functions and demonstrated a number of their properties. As well as the papers [63, 102–104], where many researchers managed to generalize various fractional integrals and derivatives.
- The papers of Sousa et al. [143–149], where the authors introduced another so-called ψ-Hilfer fractional derivative with respect to a given function and presented some important properties concerning this type of fractional operator.
- The paper of Almalahi et al. [31], which deals with the boundary value problem of ψ-Hilfer fractional derivative of the form:

© The Author(s), under exclusive license to Springer Nature Switzerland AG 2023
M. Benchohra et al., *Advanced Topics in Fractional Differential Equations*,
Synthesis Lectures on Mathematics & Statistics,
https://doi.org/10.1007/978-3-031-26928-8_3

$$\begin{cases} {}^H\mathcal{D}_{a^+}^{\alpha,\beta;\psi} x(t) = f\left(t, x(t), \int_a^t k(t,s)x(s)ds\right), \ t \in J := (a,b] \\ \mathcal{J}_{a^+}^{1-\gamma,\psi} \left[px\left(a^+\right) + qx\left(b^-\right)\right] = c, \qquad\qquad (\gamma = \alpha + \beta - \alpha\beta) \end{cases}$$

where ${}^H\mathcal{D}_{a^+}^{\alpha,\beta;\psi}(\cdot)$ is the generalized Hilfer fractional derivative of order $\alpha \in (0,1)$ and type $\beta \in [0,1]$ and $\mathcal{J}_{a^+}^{1-\gamma,\psi}(\cdot)$ is the generalized fractional integral in the sense of Riemann-Liouville of order $1-\gamma$, and $f : J \times E \times E \to E$ is a continuous function in an abstract Banach space E, $c_1, c_2 \in \mathbb{R}$, $c_3 \in E$, $c_1 + c_2 \neq 0$, and $\int_a^t k(t,s)x(s)ds$ is a linear integral operator with $k : J \times J \to \mathbb{R}$. They discussed the E_α-Ulam-Hyers stability and the continuous dependence of the problem.

- The paper of Liu et al. [94], where they considered the ψ-Hilfer fractional differential equation:

$$\begin{cases} {}^H\mathcal{D}_{0^+}^{\alpha,\beta;\psi} x(\tau) = f(\tau, x(\tau), x(g(\tau))), \quad \tau \in J = (0,d], \\ \mathcal{J}_{0^+}^{1-\gamma;\psi} x\left(0^+\right) = c_0 \in \mathbb{R}, \\ x(\tau) = \varphi(\tau), \quad \tau \in [-h, 0], \end{cases}$$

where ${}^H\mathcal{D}_{0^+}^{\alpha,\beta;\psi}(\cdot)$ is the ψ-Hilfer fractional derivative of order $0 < \alpha \le 1$ and type $0 \le \beta \le 1$, $\mathcal{J}_{0^+}^{1-\gamma;\psi}(\cdot)$ is the Riemann-Liouville fractional integral of order $1-\gamma$, $\gamma = \alpha + \beta(1-\alpha)$ with respect to the function ψ, and $f : J \times \mathbb{R} \times \mathbb{R} \to \mathbb{R}$ is a given function. They established the existence and uniqueness of solutions to the problem and introduced the Ulam-Hyers-Mittag-Leffler stability of the solutions.

3.2   Existence and Ulam Stability Results for Generalized Hilfer-Type Boundary Value Problem

In this section, we establish the existence and Ulam stability results for the boundary value problem of the following generalized Hilfer-type fractional differential equation:

$$\left({}^\rho\mathcal{D}_{a^+}^{\alpha,\beta} u\right)(t) = f\left(t, u(t), \left({}^\rho\mathcal{D}_{a^+}^{\alpha,\beta} u\right)(t)\right), \text{ for each }, \ t \in J, \qquad (3.1)$$

$$l\left({}^\rho\mathcal{J}_{a^+}^{1-\gamma} u\right)(a^+) + m\left({}^\rho\mathcal{J}_{a^+}^{1-\gamma} u\right)(b) = \phi, \qquad (3.2)$$

where ${}^\rho\mathcal{D}_{a^+}^{\alpha,\beta}, {}^\rho\mathcal{J}_{a^+}^{1-\gamma}$ are the generalized Hilfer-type fractional derivative of order $\alpha \in (0,1)$ and type $\beta \in [0,1]$ and generalized fractional integral of order $1-\gamma$, $(\gamma = \alpha + \beta - \alpha\beta)$, respectively, $\phi \in E$, $f : J \times E \times E \to E$ is a given function, and l, m are reals with $l + m \neq 0$.

3.2.1 Existence Results

We consider the following linear fractional differential equation:

$$\left({}^{\rho}\mathcal{D}_{a+}^{\alpha,\beta} u \right)(t) = \psi(t), \quad t \in J, \tag{3.3}$$

where $0 < \alpha < 1, 0 \le \beta \le 1, \rho > 0$, with the boundary condition

$$l \left({}^{\rho}\mathcal{J}_{a+}^{1-\gamma} u \right)(a^+) + m \left({}^{\rho}\mathcal{J}_{a+}^{1-\gamma} u \right)(b) = \phi, \tag{3.4}$$

where $\gamma = \alpha + \beta - \alpha\beta, \phi \in E$, and $l, m \in \mathbb{R}$ with $l + m \ne 0$. The following theorem shows that the problem (3.3)–(3.4) has a unique solution given by

$$u(t) = \left[\phi - \frac{m}{\Gamma(1-\gamma+\alpha)} \int_a^b \left(\frac{b^\rho - s^\rho}{\rho} \right)^{\alpha-\gamma} s^{\rho-1} \psi(s)\,ds \right]$$

$$\times \frac{1}{(l+m)\Gamma(\gamma)} \left(\frac{t^\rho - a^\rho}{\rho} \right)^{\gamma-1} + \frac{1}{\Gamma(\alpha)} \int_a^t \left(\frac{t^\rho - s^\rho}{\rho} \right)^{\alpha-1} s^{\rho-1} \psi(s)\,ds. \tag{3.5}$$

Theorem 3.1 *Let* $\gamma = \alpha + \beta - \alpha\beta$, *where* $0 < \alpha < 1$ *and* $0 \le \beta \le 1 \rho > 0$. *If* $\psi : J \to E$ *is a function such that* $\psi(\cdot) \in C_{\gamma,\rho}(J)$, *then* $u \in C_{\gamma,\rho}^\gamma(J)$ *satisfies the problem* (3.3)–(3.4) *if and only if it satisfies* (3.5).

Proof By Lemma 2.38, the solution of (3.3) can be written as

$$u(t) = \frac{\left({}^{\rho}\mathcal{J}_{a+}^{1-\gamma} u \right)(a^+)}{\Gamma(\gamma)} \left(\frac{t^\rho - a^\rho}{\rho} \right)^{\gamma-1} + \frac{1}{\Gamma(\alpha)} \int_a^t \left(\frac{t^\rho - s^\rho}{\rho} \right)^{\alpha-1} s^{\rho-1} \psi(s)\,ds. \tag{3.6}$$

Applying ${}^{\rho}\mathcal{J}_{a+}^{1-\gamma}$ on both sides of (3.6), using Lemma 2.19, and taking $t = b$, we obtain

$$\left({}^{\rho}\mathcal{J}_{a+}^{1-\gamma} u \right)(b) = \left({}^{\rho}\mathcal{J}_{a+}^{1-\gamma} u \right)(a^+) + \frac{1}{\Gamma(1-\gamma+\alpha)} \int_a^b \left(\frac{b^\rho - s^\rho}{\rho} \right)^{\alpha-\gamma} s^{\rho-1} \psi(s)\,ds, \tag{3.7}$$

multiplying both sides of (3.7) by m, we get

$$m \left({}^{\rho}\mathcal{J}_{a+}^{1-\gamma} u \right)(b) = m \left({}^{\rho}\mathcal{J}_{a+}^{1-\gamma} u \right)(a^+)$$

$$+ \frac{m}{\Gamma(1-\gamma+\alpha)} \int_a^b \left(\frac{b^\rho - s^\rho}{\rho} \right)^{\alpha-\gamma} s^{\rho-1} \psi(s)\,ds.$$

Using condition (3.4), we obtain

$$m \left({}^{\rho}\mathcal{J}_{a+}^{1-\gamma} u \right)(b) = \phi - l \left({}^{\rho}\mathcal{J}_{a+}^{1-\gamma} u \right)(a^+).$$

Thus

$$\phi - l \left({}^{\rho}\mathcal{J}_{a+}^{1-\gamma} u \right)(a^+) = m \left({}^{\rho}\mathcal{J}_{a+}^{1-\gamma} u \right)(a^+)$$

$$+ \frac{m}{\Gamma(1-\gamma+\alpha)} \int_a^b \left(\frac{b^\rho - s^\rho}{\rho} \right)^{\alpha-\gamma} s^{\rho-1} \psi(s) ds,$$

which implies that

$$\left({}^{\rho}\mathcal{J}_{a+}^{1-\gamma} u \right)(a^+) = \frac{\phi}{l+m} - \frac{m \int_a^b \left(\frac{b^\rho - s^\rho}{\rho} \right)^{\alpha-\gamma} s^{\rho-1} \psi(s) ds}{(l+m)\Gamma(1-\gamma+\alpha)}. \tag{3.8}$$

Substituting (3.8) into (3.6), we obtain (3.5).

Reciprocally, applying ${}^{\rho}\mathcal{J}_{a+}^{1-\gamma}$ on both sides of (3.5), using Lemma 2.19 and Theorem 2.14, we get

$$\left({}^{\rho}\mathcal{J}_{a+}^{1-\gamma} u \right)(t) = \frac{\phi}{l+m} - \frac{m}{(l+m)} \left({}^{\rho}\mathcal{J}_{a+}^{1-\gamma+\alpha} \psi \right)(b) + \left({}^{\rho}\mathcal{J}_{a+}^{1-\gamma+\alpha} \psi \right)(t). \tag{3.9}$$

Next, taking the limit $t \to a^+$ of (3.9) and using Lemma 2.24, with $1 - \gamma < 1 - \gamma + \alpha$, we obtain

$$\left({}^{\rho}\mathcal{J}_{a+}^{1-\gamma} u \right)(a^+) = \frac{\phi}{l+m} - \frac{m}{(l+m)} \left({}^{\rho}\mathcal{J}_{a+}^{1-\gamma+\alpha} \psi \right)(b). \tag{3.10}$$

Now, taking $t = b$ in (3.9), we get

$$\left({}^{\rho}\mathcal{J}_{a+}^{1-\gamma} u \right)(b) = \frac{\phi}{l+m} - \frac{m}{(l+m)} \left({}^{\rho}\mathcal{J}_{a+}^{1-\gamma+\alpha} \psi \right)(b) + \left({}^{\rho}\mathcal{J}_{a+}^{1-\gamma+\alpha} \psi \right)(b). \tag{3.11}$$

From (3.10) and (3.11), we find that

$$l \left({}^{\rho}\mathcal{J}_{a+}^{1-\gamma} u \right)(a^+) + m \left({}^{\rho}\mathcal{J}_{a+}^{1-\gamma} u \right)(b) = \frac{l.\phi}{l+m} - \frac{lm}{l+m} \left({}^{\rho}\mathcal{J}_{a+}^{1-\gamma+\alpha} \psi \right)(b)$$

$$+ \frac{m.\phi}{l+m} - \frac{m^2}{l+m} \left({}^{\rho}\mathcal{J}_{a+}^{1-\gamma+\alpha} \psi \right)(b)$$

$$+ m \left({}^{\rho}\mathcal{J}_{a+}^{1-\gamma+\alpha} \psi \right)(b)$$

$$= \phi + \left(m - \frac{lm - m^2}{l+m} \right) \left({}^{\rho}\mathcal{J}_{a+}^{1-\gamma+\alpha} \psi \right)(b)$$

$$= \phi,$$

which shows that the boundary condition $l \left({}^{\rho}\mathcal{J}_{a+}^{1-\gamma} u \right) (a^+) + m \left({}^{\rho}\mathcal{J}_{a+}^{1-\gamma} u \right) (b) = \phi$ is satisfied. Next, apply operator ${}^{\rho}\mathcal{D}_{a+}^{\gamma}$ on both sides of (3.5). Then, from Lemmas 2.19 and 2.33, we obtain

$$({}^{\rho}\mathcal{D}_{a+}^{\gamma} u)(t) = \left({}^{\rho}\mathcal{D}_{a+}^{\beta(1-\alpha)} \psi \right)(t). \tag{3.12}$$

Since $u \in C_{\gamma,\rho}^{\gamma}(J)$ and by definition of $C_{\gamma,\rho}^{\gamma}(J)$, we have ${}^{\rho}\mathcal{D}_{a+}^{\gamma} u \in C_{\gamma,\rho}(J)$, then (3.12) implies that

$$({}^{\rho}\mathcal{D}_{a+}^{\gamma} u)(t) = \left(\delta_{\rho} \, {}^{\rho}\mathcal{J}_{a+}^{1-\beta(1-\alpha)} \psi \right)(t) = \left({}^{\rho}\mathcal{D}_{a+}^{\beta(1-\alpha)} \psi \right)(t) \in C_{\gamma,\rho}(J). \tag{3.13}$$

As $\psi(\cdot) \in C_{\gamma,\rho}(J)$ and from Lemma 2.23, it follows that

$$\left({}^{\rho}\mathcal{J}_{a+}^{1-\beta(1-\alpha)} \psi \right) \in C_{\gamma,\rho}(J). \tag{3.14}$$

From (3.13), (3.14), and by the Definition of the space $C_{\gamma,\rho}^{n}(J)$, we obtain

$$\left({}^{\rho}\mathcal{J}_{a+}^{1-\beta(1-\alpha)} \psi \right) \in C_{\gamma,\rho}^{1}(J).$$

Applying operator ${}^{\rho}\mathcal{J}_{a+}^{\beta(1-\alpha)}$ to both sides of (3.12) and using Lemmas 2.32 and 2.24 and Property 2.22, we have

$$\left({}^{\rho}\mathcal{D}_{a+}^{\alpha,\beta} u \right)(t) = {}^{\rho}\mathcal{J}_{a+}^{\beta(1-\alpha)} \left({}^{\rho}\mathcal{D}_{a+}^{\gamma} u \right)(t)$$

$$= \psi(t) + \frac{\left({}^{\rho}\mathcal{J}_{a+}^{1-\beta(1-\alpha)} \psi(t) \right)(a)}{\Gamma(\beta(1-\alpha))} \left(\frac{t^{\rho} - a^{\rho}}{\rho} \right)^{\beta(1-\alpha)-1}$$

$$= \psi(t),$$

that is, (3.3) holds. This completes the proof.   □

As a consequence of Theorem 3.1, we have the following result.

Lemma 3.2 *Let* $\gamma = \alpha + \beta - \alpha\beta$ *where* $0 < \alpha < 1$ *and* $0 \le \beta \le 1$, *let* $f : J \times E \times E \rightarrow E$ *be a function such that* $f(\cdot, u(\cdot), v(\cdot)) \in C_{\gamma,\rho}(J)$ *for any* $u, v \in C_{\gamma,\rho}(J)$. *If* $u \in C_{\gamma,\rho}^{\gamma}(J)$, *then* u *satisfies the problem* (3.1)–(3.2) *if and only if* u *is the fixed point of the operator* $\Psi : C_{\gamma,\rho}(J) \rightarrow C_{\gamma,\rho}(J)$ *defined by*

$$\Psi u(t) = \left[\phi - \frac{m}{\Gamma(1-\gamma+\alpha)} \int_{a}^{b} \left(\frac{b^{\rho} - s^{\rho}}{\rho} \right)^{\alpha-\gamma} s^{\rho-1} h(s) ds \right]$$

$$\times \frac{1}{(l+m)\Gamma(\gamma)} \left(\frac{t^{\rho} - a^{\rho}}{\rho} \right)^{\gamma-1} + \frac{1}{\Gamma(\alpha)} \int_{a}^{t} \left(\frac{t^{\rho} - s^{\rho}}{\rho} \right)^{\alpha-1} s^{\rho-1} h(s) ds, \tag{3.15}$$

where $h : J \rightarrow E$ *be a function satisfying the functional equation*

$$h(t) = f(t, u(t), h(t)).$$

Clearly, $h \in C_{\gamma,\rho}(J)$. *Also, by Lemma 2.23,* $\Psi u \in C_{\gamma,\rho}(J)$.

Lemma 3.3 ([75]) *Let* $D \subset C_{\gamma,\rho}(J)$ *be a bounded and equicontinuous set, then*

(i) the function $t \rightarrow \mu\left(\left(\dfrac{t^\rho - a^\rho}{\rho}\right)^{1-\gamma} D(t)\right)$ *is continuous on* $\bar{J}$, *and*

$$\mu_{C_{\gamma,\rho}}(D) = \sup_{t \in \bar{J}} \mu\left(\left(\frac{t^\rho - a^\rho}{\rho}\right)^{1-\gamma} D(t)\right).$$

(ii) $\mu\left(\left\{\int_a^b u(s)ds : u \in D\right\}\right) \leq \int_a^b \mu(D(s))ds$, *where*

$$D(t) = \{u(t) : u \in D\}, t \in J.$$

We are now in a position to state and prove our existence result for the problem (3.1)–(3.2) based on Theorem 2.49.

Theorem 3.4 *Assume that the following hypotheses hold:*

(3.4.1) *The function* $t \mapsto f(t, u, v)$ *is measurable and continuous on* J *for each* $u, v \in E$, *and the functions* $u \mapsto f(t, u, v)$ *and* $v \mapsto f(t, u, v)$ *are continuous on* E *for a.e.* $t \in J$.

(3.4.2) *There exists a continuous function* $p : \bar{J} \longrightarrow [0, \infty)$ *such that*

$$\|f(t, u, v)\| \leq p(t), \text{ for a.e. } t \in J \text{ and for each } u, v \in E.$$

(3.4.3) *For each bounded set* $B \subset E$ *and for each* $t \in J$, *we have*

$$\mu(f(t, B, (^\rho D_{a^+}^{\alpha,\beta} B))) \leq \left(\frac{t^\rho - a^\rho}{\rho}\right)^{1-\gamma} p(t)\mu(B),$$

where $^\rho D_{a^+}^{\alpha,\beta} B = \{^\rho D_{a^+}^{\alpha,\beta} w : w \in B\}$ *and* $p^* = \sup_{t \in \bar{J}} p(t)$.

If

$$\ell := \frac{p^*}{\Gamma(\alpha + 1)} \left(\frac{b^\rho - a^\rho}{\rho}\right)^{1-\gamma+\alpha} < 1, \qquad (3.16)$$

then the problem (3.1)–(3.2) has at least one solution defined on J.

Proof Consider the operator $\Psi : C_{\gamma,\rho}(J) \to C_{\gamma,\rho}(J)$ defined in (3.15).
For any $u \in C_{\gamma,\rho}(J)$, and each $t \in J$ we have

$$\left\| \left(\frac{t^\rho - a^\rho}{\rho} \right)^{1-\gamma} (\Psi u)(t) \right\| \leq \frac{\|\phi\|}{|l + m|\Gamma(\gamma)} + \frac{|m| \int_a^b \left(\frac{b^\rho - s^\rho}{\rho} \right)^{\alpha-\gamma} s^{\rho-1} \|h(s)\| ds}{|l + m|\Gamma(\gamma)\Gamma(1 - \gamma + \alpha)}$$

$$+ \left(\frac{t^\rho - a^\rho}{\rho} \right)^{1-\gamma} \int_a^t \left(\frac{t^\rho - s^\rho}{\rho} \right)^{\alpha-1} s^{\rho-1} \frac{\|h(s)\|}{\Gamma(\alpha)} ds$$

$$\leq \frac{\|\phi\|}{|l + m|\Gamma(\gamma)} + \frac{|m| p^*}{|l + m|\Gamma(\gamma)} \left({}^\rho \mathcal{J}_{a^+}^{1-\gamma+\alpha}(1) \right)(b)$$

$$+ p^* \left(\frac{t^\rho - a^\rho}{\rho} \right)^{1-\gamma} \left({}^\rho \mathcal{J}_{a^+}^{\alpha}(1) \right)(t).$$

By Lemma 2.19, we have

$$\left\| \left(\frac{t^\rho - a^\rho}{\rho} \right)^{1-\gamma} (\Psi u)(t) \right\|$$

$$\leq \frac{\|\phi\|}{|l + m|\Gamma(\gamma)} + \frac{|m| p^*}{|l + m|\Gamma(\gamma)\Gamma(\alpha - \gamma)} \left(\frac{b^\rho - a^\rho}{\rho} \right)^{1-\gamma+\alpha}$$

$$+ \frac{p^*}{\Gamma(\alpha + 1)} \left(\frac{t^\rho - a^\rho}{\rho} \right)^{1-\gamma+\alpha}.$$

Hence, for any $u \in C_{\gamma,\rho}(J)$, and each $t \in J$, we get

$$\|(\Psi u)\|_{C_{\gamma,\rho}}$$

$$\leq \frac{\|\phi\|}{|l + m|\Gamma(\gamma)}$$

$$+ \frac{|m| p^*}{|l + m|\Gamma(\gamma)\Gamma(\alpha - \gamma)} \left(\frac{b^\rho - a^\rho}{\rho} \right)^{1-\gamma+\alpha} + \frac{p^*}{\Gamma(\alpha + 1)} \left(\frac{t^\rho - a^\rho}{\rho} \right)^{1-\gamma+\alpha}$$

$$:= R.$$

This proves that Ψ transforms the ball $B_R := B(0, R) = \{w \in C_{\gamma,\rho} : \|w\|_{C_{\gamma,\rho}} \leq R\}$ into itself. We shall show that the operator $\Psi : B_R \to B_R$ satisfies all the assumptions of Theorem 2.49. The proof will be given in several steps.

Step 1: $\Psi : B_R \to B_R$ is continuous.
Let $\{u_n\}_{n\in\mathbb{N}}$ be a sequence such that $u_n \longrightarrow u$ in B_R. Then, for each $t \in J$, we have

$$\left\| \left(\frac{t^\rho - a^\rho}{\rho} \right)^{1-\gamma} (\Psi u_n)(t) - \left(\frac{t^\rho - a^\rho}{\rho} \right)^{1-\gamma} (\Psi u)(t) \right\|$$

$$\leq \frac{|m|}{|l + m| \Gamma(\gamma) \Gamma(1 - \gamma + \alpha)} \int_a^b \left(\frac{b^\rho - s^\rho}{\rho} \right)^{\alpha - \gamma} s^{\rho - 1} \|h_n(s) - h(s)\| ds \qquad (3.17)$$

$$+ \frac{1}{\Gamma(\alpha)} \left(\frac{t^\rho - a^\rho}{\rho} \right)^{1-\gamma} \int_a^t \left(\frac{t^\rho - s^\rho}{\rho} \right)^{\alpha - 1} s^{\rho - 1} \|h_n(s) - h(s)\| ds,$$

where $h_n, h \in C_{\gamma, \rho}(J)$ be such that

$$h_n(t) = f(t, u_n(t), h_n(t)), \ h(t) = f(t, u(t), h(t)).$$

Since $u_n \longrightarrow u$ as $n \longrightarrow \infty$ and f is continuous, then by the Lebesgue dominated convergence theorem, Eq. (3.17) implies

$$\|\Psi u_n - \Psi u\|_{C_{\gamma, \rho}} \longrightarrow 0 as n \longrightarrow \infty.$$

Step 2: $\Psi(B_R)$ is bounded and equicontinuous.
Since $\Psi(B_R) \subset B_R$ and B_R is bounded, then $\Psi(B_R)$ is bounded.

Next, let $t_1, t_2 \in J$ such that $a < t_1 < t_2 \leq b$ and let $u \in B_R$. Thus, we have

$$\left\| \left(\frac{t_2^\rho - a^\rho}{\rho} \right)^{1-\gamma} (\Psi u)(t_2) - \left(\frac{t_1^\rho - a^\rho}{\rho} \right)^{1-\gamma} (\Psi u)(t_1) \right\|$$

$$\leq \left\| \frac{1}{\Gamma(\alpha)} \left(\frac{t_2^\rho - a^\rho}{\rho} \right)^{1-\gamma} \int_a^{t_2} \left(\frac{t_2^\rho - s^\rho}{\rho} \right)^{\alpha - 1} s^{\rho - 1} h(s) ds \right.$$

$$\left. - \frac{1}{\Gamma(\alpha)} \left(\frac{t_1^\rho - a^\rho}{\rho} \right)^{1-\gamma} \int_a^{t_1} \left(\frac{t_1^\rho - s^\rho}{\rho} \right)^{\alpha - 1} s^{\rho - 1} h(s) ds \right\|,$$

then,

$$\left\| \left(\frac{t_2^\rho - a^\rho}{\rho} \right)^{1-\gamma} (\Psi u)(t_2) - \left(\frac{t_1^\rho - a^\rho}{\rho} \right)^{1-\gamma} (\Psi u)(t_1) \right\|$$

$$\leq \frac{1}{\Gamma(\alpha)} \left(\frac{t_2^\rho - a^\rho}{\rho} \right)^{1-\gamma} \int_{t_1}^{t_2} \left(\frac{t_2^\rho - s^\rho}{\rho} \right)^{\alpha - 1} s^{\rho - 1} \|h(s)\| ds + \rho^{\alpha - \gamma}$$

$$\times \int_a^{t_1} \left| (t_2^\rho - a^\rho)^{1-\gamma} (t_2^\rho - s^\rho)^{\alpha - 1} - (t_1^\rho - a^\rho)^{1-\gamma} (t_1^\rho - s^\rho)^{\alpha - 1} \right| \frac{s^{\rho - 1} \|h(s)\|}{\Gamma(\alpha)} ds$$

$$\leq p^* \left(\frac{b^\rho - a^\rho}{\rho} \right)^{1-\gamma} \left({}^\rho \mathcal{J}_{t_1^+}^\alpha (1) \right)(t_2) + \rho^{\alpha - \gamma}$$

$$\times \int_a^{t_1} \left| (t_2^\rho - a^\rho)^{1-\gamma} (t_2^\rho - s^\rho)^{\alpha - 1} - (t_1^\rho - a^\rho)^{1-\gamma} (t_1^\rho - s^\rho)^{\alpha - 1} \right| \frac{p^* s^{\rho - 1} ds}{\Gamma(\alpha)}.$$

By Lemma 2.19, we have

$$\left\| \left(\frac{t_2^\rho - a^\rho}{\rho} \right)^{1-\gamma} (\Psi u)(t_2) - \left(\frac{t_1^\rho - a^\rho}{\rho} \right)^{1-\gamma} (\Psi u)(t_1) \right\|$$

$$\leq \frac{p^*}{\Gamma(\alpha+1)} \left(\frac{b^\rho - a^\rho}{\rho} \right)^{1-\gamma} \left(\frac{t_2^\rho - t_1^\rho}{\rho} \right)^\alpha + \rho^{\alpha-\gamma}$$

$$+ \int_a^{t_1} \left| (t_2^\rho - a^\rho)^{1-\gamma} (t_2^\rho - s^\rho)^{\alpha-1} - (t_1^\rho - a^\rho)^{1-\gamma} (t_1^\rho - s^\rho)^{\alpha-1} \right| \frac{p^* s^{\rho-1} ds}{\Gamma(\alpha)}.$$

As $t_1 \longrightarrow t_2$, the right side of the above inequality tends to zero. Hence, $\Psi(B_R)$ is bounded and equicontinuous.

Step 3: The implication (2.11) of Theorem 2.49 holds.

Now let D be an equicontinuous subset of B_R such that $D \subset \overline{\Psi(D)} \cup \{0\}$; therefore, the function $t \longrightarrow d(t) = \mu(D(t))$ is continuous on J. By hypothesis (3.4.3) and the properties of the measure μ, for each $t \in J$, we have

$$\left(\frac{t^\rho - a^\rho}{\rho} \right)^{1-\gamma} d(t) \leq \mu \left(\left(\frac{t^\rho - a^\rho}{\rho} \right)^{1-\gamma} (\Psi D)(t) \cup \{0\} \right)$$

$$\leq \mu \left(\left(\frac{t^\rho - a^\rho}{\rho} \right)^{1-\gamma} (\Psi D)(t) \right)$$

$$\leq \left(\frac{b^\rho - a^\rho}{\rho} \right)^{1-\gamma}$$

$$\times \int_a^t \left(\frac{t^\rho - s^\rho}{\rho} \right)^{\alpha-1} s^{\rho-1} \frac{p(s)\mu(D(s))}{\Gamma(\alpha)} \left(\frac{s^\rho - a^\rho}{\rho} \right)^{1-\gamma} ds$$

$$\leq p^* \left(\frac{b^\rho - a^\rho}{\rho} \right)^{1-\gamma} \|d\|_{C_{\gamma,\rho}} \left({}^\rho \mathcal{J}_{a^+}^\alpha (1) \right)(t)$$

$$\leq \frac{p^*}{\Gamma(\alpha+1)} \left(\frac{b^\rho - a^\rho}{\rho} \right)^{1-\gamma+\alpha} \|d\|_{C_{\gamma,\rho}}.$$

Thus

$$\|d\|_{C_{\gamma,\rho}} \leq \ell \|d\|_{C_{\gamma,\rho}}.$$

From (3.16), we get $\|d\|_{C_{\gamma,\rho}} = 0$, that is $d(t) = \mu(D(t)) = 0$, for each $t \in J$, and then $D(t)$ is relatively compact in E. In view of the Ascoli-Arzela Theorem, D is relatively compact in B_R. Applying now Theorem 2.49, we conclude that Ψ has a fixed point, which is solution of the problem (3.1)–(3.2). □

Our next existence result for the problem (3.1)–(3.2) is based on Theorem 2.48 (Darbo's fixed point theorem).

Theorem 3.5 *Assume that the hypotheses (3.4.1)–(3.4.3) and the condition (3.16) hold. Then the problem (3.1)–(3.2) has a solution defined on J.*

Proof Consider the operator Ψ defined in (3.15). We know that $\Psi : B_R \longrightarrow B_R$ is bounded and continuous and that $\Psi(B_R)$ is equicontinuous, we need to prove that the operator Ψ is a ℓ-contraction.

Let $D \subset B_R$ and $t \in J$. Then we have

$$\mu\left(\left(\frac{t^\rho - a^\rho}{\rho}\right)^{1-\gamma} (\Psi D)(t)\right) = \mu\left(\left(\frac{t^\rho - a^\rho}{\rho}\right)^{1-\gamma} (\Psi u)(t) : u \in D\right)$$

$$\leq \left\{\int_a^t \left(\frac{t^\rho - s^\rho}{\rho}\right)^{\alpha-1} \frac{s^{\rho-1} p(s)\mu(D(s))}{\Gamma(\alpha)} \left(\frac{s^\rho - a^\rho}{\rho}\right)^{1-\gamma} ds : u \in D\right\}$$

$$\times \left(\frac{b^\rho - a^\rho}{\rho}\right)^{1-\gamma}$$

$$\leq p^* \left(\frac{b^\rho - a^\rho}{\rho}\right)^{1-\gamma} \mu_{C_{\gamma,\rho}}(D) \left({}^\rho\mathcal{J}_{a^+}^\alpha (1)\right)(t)$$

$$\leq \frac{p^*}{\Gamma(\alpha+1)} \left(\frac{b^\rho - a^\rho}{\rho}\right)^{1-\gamma+\alpha} \mu_{C_{\gamma,\rho}}(D).$$

Therefore,

$$\mu_{C_{\gamma,\rho}}(\Psi D) \leq \frac{p^*}{\Gamma(\alpha+1)} \left(\frac{b^\rho - a^\rho}{\rho}\right)^{1-\gamma+\alpha} \mu_{C_{\gamma,\rho}}(D).$$

So, by (3.16), the operator Ψ is a ℓ-contraction, where

$$\ell := \frac{p^*}{\Gamma(\alpha+1)} \left(\frac{b^\rho - a^\rho}{\rho}\right)^{1-\gamma+\alpha} < 1.$$

Consequently, from Theorem 2.48, we conclude that Ψ has a fixed point $u \in B_R$, which is a solution to problem (3.1)–(3.2). $\square$

3.2.2 Ulam-Hyers-Rassias Stability

Now we are concerned with the generalized Ulam-Hyers-Rassias stability of our Eq. (3.1). Let $\epsilon > 0$ and $\theta : J \longrightarrow [0, \infty)$ be a continuous function. We consider the following inequalities :

$$\left\|\left({}^\rho\mathcal{D}_{a^+}^{\alpha,\beta} u\right)(t) - f\left(t, u(t), \left({}^\rho\mathcal{D}_{a^+}^{\alpha,\beta} u\right)(t)\right)\right\| \leq \epsilon; \quad t \in J, \tag{3.18}$$

$$\left\|\left({}^\rho\mathcal{D}_{a^+}^{\alpha,\beta} u\right)(t) - f\left(t, u(t), \left({}^\rho\mathcal{D}_{a^+}^{\alpha,\beta} u\right)(t)\right)\right\| \leq \theta(t); \quad t \in J, \tag{3.19}$$

$$\left\|\left({}^\rho\mathcal{D}_{a^+}^{\alpha,\beta} u\right)(t) - f\left(t, u(t), \left({}^\rho\mathcal{D}_{a^+}^{\alpha,\beta} u\right)(t)\right)\right\| \leq \epsilon\theta(t); \quad t \in J. \tag{3.20}$$

Definition 3.6 ([51, 52]) Problem (3.1)–(3.2) is Ulam-Hyers (U-H) stable if there exists a real number $a_f > 0$ such that for each $\epsilon > 0$ and for each solution $u \in C_{\gamma,\rho}(J)$ of inequality (3.18) there exists a solution $v \in C_{\gamma,\rho}(J)$ of (3.1)–(3.2) with

$$\|u(t) - v(t)\| \leq \epsilon a_f; \quad t \in J.$$

Definition 3.7 ([51, 52]) Problem (3.1)–(3.2) is generalized Ulam-Hyers (G.U-H) stable if there exists $a_f : C([0, \infty), [0, \infty))$ with $a_f(0) = 0$ such that for each $\epsilon > 0$ and for each solution $u \in C_{\gamma,\rho}(J)$ of inequality (3.18) there exists a solution $v \in C_{\gamma,\rho}(J)$ of (3.1)–(3.2) with

$$\|u(t) - v(t)\| \leq a_f(\epsilon); \quad t \in J.$$

Definition 3.8 ([51, 52]) Problem (3.1)–(3.2) is Ulam-Hyers-Rassias (U-H-R) stable with respect to θ if there exists a real number $a_{f,\theta} > 0$ such that for each $\epsilon > 0$ and for each solution $u \in C_{\gamma,\rho}(J)$ of inequality (3.20) there exists a solution $v \in C_{\gamma,\rho}(J)$ of (3.1)–(3.2) with

$$\|u(t) - v(t)\| \leq \epsilon a_{f,\theta}\theta(t); \quad t \in J.$$

Definition 3.9 ([51, 52]) Problem (3.1)–(3.2) is generalized Ulam-Hyers-Rassias (G.U-H-R) stable with respect to θ if there exists a real number $a_{f,\theta} > 0$ such that for each solution $u \in C_{\gamma,\rho}(J)$ of inequality (3.19) there exists a solution $v \in C_{\gamma,\rho}(J)$ of (3.1)–(3.2) with

$$\|u(t) - v(t)\| \leq a_{f,\theta}\theta(t); \quad t \in J.$$

Remark 3.10 It is clear that

1. Definition 3.6 $\Longrightarrow$ Definition 3.7.
2. Definition 3.8 $\Longrightarrow$ Definition 3.9.
3. Definition 3.8 for $\theta(.) = 1 \Longrightarrow$ Definition 3.6.

Theorem 3.11 *Assume that the hypotheses (3.4.1), (3.4.2), and the following hypotheses hold:*

(3.11.1) *There exists $\lambda_\theta > 0$ such that for each $t \in J$, we have*

$$(^\rho J_{a^+}^\alpha \theta)(t) \leq \lambda_\theta \theta(t).$$

(3.11.2) *There exists a continuous function $q : \bar{J} \longrightarrow [0, \infty)$ such that for each $t \in J$, we have*

$$p(t) \leq q(t)\theta(t).$$

Then equation Problem (3.1)–(3.2) is G.U-H-R stable.

Proof Consider the operator Ψ defined in (3.15). Let u be a solution of inequality (3.19), and let us assume that v is a solution of the problem (3.1)–(3.2). Thus, we have

$$\Psi v(t) = \left[\phi - \frac{m}{\Gamma(1 - \gamma + \alpha)} \int_a^b \left(\frac{b^\rho - s^\rho}{\rho} \right)^{\alpha - \gamma} s^{\rho - 1} g(s) ds \right]$$

$$\times \frac{1}{(l + m)\Gamma(\gamma)} \left(\frac{t^\rho - a^\rho}{\rho} \right)^{\gamma - 1}$$

$$+ \frac{1}{\Gamma(\alpha)} \int_a^t \left(\frac{t^\rho - s^\rho}{\rho} \right)^{\alpha - 1} s^{\rho - 1} g(s) ds, \ t \in J,$$

where $g : J \to E$ be a function satisfying

$$g(t) = f(t, v(t), g(t)).$$

From inequality (3.19), for each $t \in J$, we have

$$\left\| u(t) - \left[\phi - \frac{m}{\Gamma(1 - \gamma + \alpha)} \int_a^b \left(\frac{b^\rho - s^\rho}{\rho} \right)^{\alpha - \gamma} s^{\rho - 1} h(s) ds \right] \right.$$

$$\times \frac{1}{(l + m)\Gamma(\gamma)} \left(\frac{t^\rho - a^\rho}{\rho} \right)^{\gamma - 1} - \frac{1}{\Gamma(\alpha)} \int_a^t \left(\frac{t^\rho - s^\rho}{\rho} \right)^{\alpha - 1} s^{\rho - 1} h(s) ds \left\| \right.$$

$$\leq (^\rho \mathcal{J}_{a^+}^\alpha \theta)(t).$$

Set $q^* = \sup_{t \in J} q(t)$.

From hypotheses (3.11.1) and (3.11.2), for each $t \in J$, we get

$$\| u(t) - v(t) \| \leq \left\| u(t) - \left[\phi - \frac{m}{\Gamma(1 - \gamma + \alpha)} \int_a^b \left(\frac{b^\rho - s^\rho}{\rho} \right)^{\alpha - \gamma} s^{\rho - 1} h(s) ds \right] \right.$$

$$\times \frac{\left(\frac{t^\rho - a^\rho}{\rho} \right)^{\gamma - 1}}{(l + m)\Gamma(\gamma)} - \frac{1}{\Gamma(\alpha)} \int_a^t \left(\frac{t^\rho - s^\rho}{\rho} \right)^{\alpha - 1} s^{\rho - 1} h(s) ds \left\| \right.$$

$$+ \frac{1}{\Gamma(\alpha)} \int_a^t \left(\frac{t^\rho - s^\rho}{\rho} \right)^{\alpha - 1} s^{\rho - 1} \| h(s) - g(s) \| ds$$

$$\leq (^\rho \mathcal{J}_{a^+}^\alpha \theta)(t) + \frac{1}{\Gamma(\alpha)} \int_a^t \left(\frac{t^\rho - s^\rho}{\rho} \right)^{\alpha - 1} s^{\rho - 1} 2q^* \theta(s) ds$$

$$\leq \lambda_\theta \theta(t) + 2q^* (^\rho \mathcal{J}_{a^+}^\alpha \theta)(t)$$

$$\leq [1 + 2q^*] \lambda_\theta \theta(t)$$

$$:= a_{f, \theta} \theta(t).$$

Hence, Eq. (3.1) is G.U-H-R stable.   □

3.2.3   An Example

Let

$$E = l^1 = \left\{ u = (u_1, u_2, \ldots, u_n, \ldots), \sum_{n=1}^{\infty} |u_n| < \infty \right\}$$

be the Banach space with the norm

$$\|u\| = \sum_{n=1}^{\infty} |u_n|.$$

Consider the following boundary value problem of fractional differential equation:

$$^1D_{1+}^{\frac{1}{2},0} u_n(t) = f_n \left(t, u_n(t), \left({}^1D_{1+}^{\frac{1}{2},0} u_n \right)(t) \right), \quad t \in (1, e] \tag{3.21}$$

$$\left({}^1\mathcal{J}_{1+}^{\frac{1}{2}} u_n \right)(1^+) + \left({}^1\mathcal{J}_{1+}^{\frac{1}{2}} u_n \right)(e) = 0, \tag{3.22}$$

where

$$f_n \left(t, u_n(t), \left({}^1D_{1+}^{\frac{1}{2},0} u_n \right)(t) \right) = \frac{ct^2}{e^2} \left(\sin(t-1) + u_n(t) \right), \quad t \in (1, e].$$

Let

$$f = (f_1, f_2, \ldots, f_n, \ldots), u = (u_1, u_2, \ldots, u_n, \ldots) \quad c = \frac{1}{4}\Gamma\left(\frac{1}{2}\right),$$

$\gamma = \alpha = \frac{1}{2}$, $\rho = 1$, and $\beta = 0$. Clearly, the function f is continuous.
    The hypothesis (3.4.2) is satisfied with

$$p(t) = \frac{ct^2 |\sin(t-1)|}{e^2}, \quad t \in (1, e].$$

A simple computation shows that the conditions of Theorem 3.4 are satisfied. Hence the problem (3.21)–(3.22) has at least one solution defined on $[1, e]$.

Also, hypotheses (3.11.1) and (3.11.2) are satisfied with $\theta(t) = e^2$, $q(t) = \dfrac{p(t)}{e^2}$, and $\lambda_\theta = \dfrac{4}{\sqrt{\pi}}$. Indeed, for each $t \in (1, e]$, we get

$$(^\rho \mathcal{J}_{a+}^\alpha \theta)(t) \le \frac{4e^2}{\sqrt{\pi}}$$
$$= \lambda_\theta \theta(t).$$

Consequently, Theorem 3.11 implies that Eq. (3.21) is G.U-H-R stable.

3.3   Existence and Ulam Stability Results for k-Generalized ψ-Hilfer Initial Value Problem

In this section, we consider the initial value problem with nonlinear implicit k-generalized ψ-Hilfer-type fractional differential equation:

$$\left({}^{H}_{k}\mathcal{D}^{\alpha,\beta;\psi}_{a+}x\right)(t) = f\left(t, x(t), \left({}^{H}_{k}\mathcal{D}^{\alpha,\beta;\psi}_{a+}x\right)(t)\right), \quad t \in J, \tag{3.23}$$

$$\left(\mathcal{J}^{k(1-\xi),k;\psi}_{a+}x\right)(a^{+}) = x_0, \tag{3.24}$$

where ${}^{H}_{k}\mathcal{D}^{\alpha,\beta;\psi}_{a+}, \mathcal{J}^{k(1-\xi),k;\psi}_{a+}$ are the k-generalized ψ-Hilfer fractional derivative of order $\alpha \in (0, 1)$ and type $\beta \in [0, 1]$, and k-generalized ψ-fractional integral of order $k(1 - \xi)$, respectively, where $\xi = \frac{1}{k}(\beta(k - \alpha) + \alpha)$, $x_0 \in \mathbb{R}, k > 0$, and $f \in C(\bar{J} \times \mathbb{R}^2, \mathbb{R})$.

3.3.1   Existence Results

We consider the following fractional differential equation:

$$\left({}^{H}_{k}\mathcal{D}^{\alpha,\beta;\psi}_{a+}x\right)(t) = w(t), \quad t \in J, \tag{3.25}$$

where $0 < \alpha < 1, 0 \leq \beta \leq 1$, with the condition

$$\left(\mathcal{J}^{k(1-\xi),k;\psi}_{a+}x\right)(a^{+}) = x_0, \tag{3.26}$$

where $\xi = \dfrac{\beta(k - \alpha) + \alpha}{k}$, $x_0 \in \mathbb{R}, k > 0$, and where $w \in C(\bar{J}, \mathbb{R})$ satisfies the functional equation:

$$w(t) = f(t, x(t), w(t)).$$

The following theorem shows that the problem (3.25)–(3.26) has a unique solution.

Theorem 3.12 *If $w(\cdot) \in C^{1}_{\xi;\psi}(J)$, then x satisfies (3.25)–(3.26) if and only if it satisfies*

$$x(t) = \frac{(\psi(t) - \psi(a))^{\xi-1}}{\Gamma_k(k\xi)}x_0 + \left(\mathcal{J}^{\alpha,k;\psi}_{a+}w\right)(t). \tag{3.27}$$

Proof Assume $x \in C^{1}_{\xi;\psi}(J)$ satisfies Eqs. (3.25) and (3.26), and applying the fractional integral operator $\mathcal{J}^{\alpha,k;\psi}_{a+}(\cdot)$ on both sides of the fractional equation (3.25), so

$$\left(\mathcal{J}^{\alpha,k;\psi}_{a+}\,{}^{H}_{k}\mathcal{D}^{\alpha,\beta;\psi}_{a+}x\right)(t) = \left(\mathcal{J}^{\alpha,k;\psi}_{a+}w\right)(t),$$

and using Theorem 2.35 and Eq. (3.26), we get

$$x(t) = \frac{(\psi(t) - \psi(a))^{\xi-1}}{\Gamma_k(k\xi)} \mathcal{J}_{a+}^{k(1-\xi),k;\psi} x(a) + \left(\mathcal{J}_{a+}^{\alpha,k;\psi} w\right)(t)$$

$$= \frac{(\psi(t) - \psi(a))^{\xi-1}}{\Gamma_k(k\xi)} x_0 + \left(\mathcal{J}_{a+}^{\alpha,k;\psi} w\right)(t).$$

Let us now prove that if x satisfies Eq. (3.27), then it satisfies Eqs. (3.25) and (3.26). Applying the fractional derivative operator ${}_k^H \mathcal{D}_{a+}^{\alpha,\beta;\psi}(\cdot)$ on both sides of the fractional equation (3.27), then we get

$$\left({}_k^H \mathcal{D}_{a+}^{\alpha,\beta;\psi} x\right)(t) = {}_k^H \mathcal{D}_{a+}^{\alpha,\beta;\psi} \left(\frac{(\psi(t) - \psi(a))^{\xi-1}}{\Gamma_k(k\xi)} x_0\right) + \left({}_k^H \mathcal{D}_{a+}^{\alpha,\beta;\psi} \mathcal{J}_{a+}^{\alpha,k;\psi} w\right)(t).$$

Using Lemmas 2.31 and 2.29, we obtain Eq. (3.25). Now we apply the operator $\mathcal{J}_{a+}^{k(1-\xi),k;\psi}(\cdot)$ on Eq. (3.27) to have

$$\left(\mathcal{J}_{a+}^{k(1-\xi),k;\psi} x\right)(t) = \frac{x_0}{\Gamma_k(k\xi)} \mathcal{J}_{a+}^{k(1-\xi),k;\psi} (\psi(t) - \psi(a))^{\xi-1}$$

$$+ \left(\mathcal{J}_{a+}^{k(1-\xi),k;\psi} \mathcal{J}_{a+}^{\alpha,k;\psi} w\right)(t).$$

Now, using Lemmas 2.18 and 2.21, we get

$$\left(\mathcal{J}_{a+}^{k(1-\xi),k;\psi} x\right)(t) = \frac{x_0}{\Gamma_k(k\xi)} \mathcal{J}_{a+}^{k(1-\xi),k;\psi} (\psi(t) - \psi(a))^{\xi-1}$$

$$+ \left(\mathcal{J}_{a+}^{k(1-\xi),k;\psi} \mathcal{J}_{a+}^{\alpha,k;\psi} w\right)(t)$$

$$= x_0 + \left(\mathcal{J}_{a+}^{k(1-\xi)+\alpha,k;\psi} w\right)(t).$$

Using Theorem 2.26 with $t \to a$, we obtain Eq. (3.26). This completes the proof.   □

As a consequence of Theorem 3.12, we have the following result.

Lemma 3.13 *Let $\xi = \dfrac{\beta(k-\alpha)+\alpha}{k}$ where $0 < \alpha < 1, 0 \le \beta \le 1$, and $k > 0$, let $f : \bar{J} \times \mathbb{R} \times \mathbb{R} \to \mathbb{R}$ be a continuous function such that $f(\cdot, x(\cdot), y(\cdot)) \in C_{\xi;\psi}^1(J)$, for any $x, y \in C_{\xi;\psi}(J)$. If $x \in C_{\xi;\psi}^1(J)$, then x satisfies the problem (3.23)–(3.24) if and only if x is the fixed point of the operator $T : C_{\xi;\psi}(J) \to C_{\xi;\psi}(J)$ defined by*

$$(Tx)(t) = \frac{(\psi(t) - \psi(a))^{\xi-1}}{\Gamma_k(k\xi)} x_0 + \frac{1}{k\Gamma_k(\alpha)} \int_a^t \frac{\psi'(s)\varphi(s)ds}{(\psi(t) - \psi(s))^{1-\frac{\alpha}{k}}}, \qquad (3.28)$$

where φ be a function satisfying the functional equation

$$\varphi(t) = f(t, x(t), \varphi(t)).$$

We are now in a position to state and prove our existence result for the problem (3.23)–(3.24) based on Banach's fixed point theorem.

Theorem 3.14 *Assume that the following hypotheses are met.*

(3.14.1) The function $f : \bar{J} \times \mathbb{R} \times \mathbb{R} \to \mathbb{R}$ is continuous and

$$f(\cdot, x(\cdot), y(\cdot)) \in C^1_{\xi;\psi}(J), \text{ for any } x, y \in C_{\xi;\psi}(J).$$

(3.14.2) There exist constants $\eta_1 > 0$ and $0 < \eta_2 < 1$ such that

$$|f(t, x, y) - f(t, \bar{x}, \bar{y})| \le \eta_1 |x - \bar{x}| + \eta_2 |y - \bar{y}|$$

for any $x, y, \bar{x}, \bar{y} \in \mathbb{R}$ and $t \in \bar{J}$.

If

$$\mathcal{L} = \frac{\eta_1 \Gamma_k(k\xi)\,(\psi(b) - \psi(a))^{\frac{\alpha}{k}}}{\Gamma_k(\alpha + k\xi)(1 - \eta_2)} < 1, \tag{3.29}$$

then the problem (3.23)–(3.24) has a unique solution in $C_{\xi;\psi}(J)$.

Proof We show that the operator $\mathcal{T}$ defined in (3.28) has a unique fixed point in $C_{\xi;\psi}(J)$.
Let $x, y \in C_{\xi;\psi}(J)$. Then, for $t \in J$ we have

$$|\mathcal{T}x(t) - \mathcal{T}y(t)| \le \frac{1}{k\Gamma_k(\alpha)} \int_a^t \frac{\psi'(s)|\varphi_1(s) - \varphi_2(s)|dt}{(\psi(t) - \psi(s))^{1 - \frac{\alpha}{k}}},$$

where φ_1 and φ_1 be functions satisfying the functional equations

$$\varphi_1(t) = f(t, x(t), \varphi_1(t)),$$
$$\varphi_2(t) = f(t, y(t), \varphi_2(t)).$$

By hypothesis (3.14.2), we have

$$|\varphi_1(t) - \varphi_2(t)| = |f(t, x(t), \varphi_1(t)) - f(t, y(t), \varphi_2(t))|$$
$$\le \eta_1 |x(t) - y(t)| + \eta_2 |\varphi_1(t) - \varphi_2(t)|.$$

Then,

$$|\varphi_1(t) - \varphi_2(t)| \le \frac{\eta_1}{1 - \eta_2} |x(t) - y(t)|.$$

Therefore, for each $t \in J$

$$|\mathcal{T}x(t) - \mathcal{T}y(t)| \leq \frac{\eta_1}{(1 - \eta_2)k\Gamma_k(\alpha)} \int_a^t \frac{\psi'(s)|x(s) - y(s)|dt}{(\psi(t) - \psi(s))^{1-\frac{\alpha}{k}}}$$

$$\leq \frac{\eta_1 \|x - y\|_{C_{\xi;\psi}}}{1 - \eta_2} \mathcal{J}_{a+}^{\alpha,k;\psi} (\psi(t) - \psi(a))^{\xi-1}.$$

By Lemma 2.21, we have

$$|\mathcal{T}x(t) - \mathcal{T}y(t)| \leq \left[\frac{\eta_1 \Gamma_k(k\xi)}{\Gamma_k(\alpha + k\xi)(1 - \eta_2)} (\psi(t) - \psi(a))^{\frac{\alpha+k\xi}{k}-1} \right] \|x - y\|_{C_{\xi;\psi}},$$

hence

$$\left| (\psi(t) - \psi(a))^{1-\xi} (\mathcal{T}x(t) - \mathcal{T}y(t)) \right| \leq \left[\frac{\eta_1 \Gamma_k(k\xi)(\psi(t) - \psi(a))^{\frac{\alpha}{k}}}{\Gamma_k(\alpha + k\xi)(1 - \eta_2)} \right] \|x - y\|_{C_{\xi;\psi}}$$

$$\leq \left[\frac{\eta_1 \Gamma_k(k\xi)(\psi(b) - \psi(a))^{\frac{\alpha}{k}}}{\Gamma_k(\alpha + k\xi)(1 - \eta_2)} \right] \|x - y\|_{C_{\xi;\psi}},$$

which implies that

$$\|\mathcal{T}x - \mathcal{T}y\|_{C_{\xi;\psi}} \leq \left[\frac{\eta_1 \Gamma_k(k\xi)(\psi(b) - \psi(a))^{\frac{\alpha}{k}}}{\Gamma_k(\alpha + k\xi)(1 - \eta_2)} \right] \|x - y\|_{C_{\xi;\psi}}.$$

By (3.29), the operator $\mathcal{T}$ is a contraction. Hence, by Banach's contraction principle, $\mathcal{T}$ has a unique fixed point $x \in C_{\xi;\psi}(J)$, which is a solution to our problem (3.23)–(3.24). $\square$

3.3.2   Ulam-Hyers-Rassias Stability

Now, we consider the Ulam stability for problem (3.23)–(3.24). Let $x \in C^1_{\xi;\psi}(J)$, $\epsilon > 0$, and $v : J \longrightarrow [0, \infty)$ be a continuous function. We consider the following inequality :

$$\left| \left({}^H_k \mathcal{D}_{a+}^{\alpha,\beta;\psi} x \right)(t) - f\left(t, x(t), \left({}^H_k \mathcal{D}_{a+}^{\alpha,\beta;\psi} x \right)(t) \right) \right| \leq \epsilon v(t), \quad t \in J. \tag{3.30}$$

Definition 3.15 Problem (3.23)–(3.24) is Ulam-Hyers-Rassias (U-H-R) stable with respect to v if there exists a real number $a_{f,v} > 0$ such that for each $\epsilon > 0$ and for each solution $x \in C^1_{\xi;\psi}(J)$ of inequality (3.30) there exists a solution $y \in C^1_{\xi;\psi}(J)$ of (3.23)–(3.24) with

$$|x(t) - y(t)| \leq \epsilon a_{f,v} v(t), \quad t \in J.$$

Remark 3.16 A function $x \in C^1_{\xi;\psi}(J)$ is a solution of inequality (3.30) if and only if there exist $\sigma \in C_{\xi;\psi}(J)$ such that

1. $|\sigma(t)| \leq \epsilon v(t), t \in J$,
2. $\left({}^H_k \mathcal{D}^{\alpha,\beta;\psi}_{a+} x \right)(t) = f\left(t, x(t), \left({}^H_k \mathcal{D}^{\alpha,\beta;\psi}_{a+} x \right)(t) \right) + \sigma(t), t \in J$.

Theorem 3.17 *Assume that in addition to assumptions (3.14.1), (3.14.2), and (3.29), the following hypothesis holds.*

(3.17.1) *There exist a nondecreasing function $v \in C^1_{\xi;\psi}(J)$ and $\kappa_v > 0$ such that for each $t \in J$, we have*

$$\left(\mathcal{J}^{\alpha,k;\psi}_{a+} v \right)(t) \leq \kappa_v v(t).$$

Then the problem (3.23)–(3.24) is U-H-R stable with respect to v.

Proof Let $x \in C^1_{\xi;\psi}(J)$ be a solution of inequality (3.30), and let us assume that y is the unique solution of the problem

$$\begin{cases} \left({}^H_k \mathcal{D}^{\alpha,\beta;\psi}_{a+} y \right)(t) = f\left(t, y(t), \left({}^H_k \mathcal{D}^{\alpha,\beta;\psi}_{a+} y \right)(t) \right); \ t \in J, \\ \left(\mathcal{J}^{k(1-\xi),k;\psi}_{a+} y \right)(a^+) = \left(\mathcal{J}^{k(1-\xi),k;\psi}_{a+} x \right)(a^+). \end{cases}$$

By Lemma 3.13, we obtain for each $t \in J$

$$y(t) = \frac{(\psi(t) - \psi(a))^{\xi-1}}{\Gamma_k(k\xi)} \mathcal{J}^{k(1-\xi),k;\psi}_{a+} y(a) + \left(\mathcal{J}^{\alpha,k;\psi}_{a+} w \right)(t),$$

where $w \in C^1_{\xi;\psi}(J)$ be a function satisfying the functional equation

$$w(t) = f(t, y(t), w(t)).$$

Since x is a solution of the inequality (3.30), by Remark 3.16, we have

$$\left({}^H_k \mathcal{D}^{\alpha,\beta;\psi}_{a+} x \right)(t) = f\left(t, x(t), \left({}^H_k \mathcal{D}^{\alpha,\beta;\psi}_{a+} x \right)(t) \right) + \sigma(t), t \in J. \qquad (3.31)$$

Clearly, the solution of (3.31) is given by

$$x(t) = \frac{(\psi(t) - \psi(a))^{\xi-1}}{\Gamma_k(k\xi)} \mathcal{J}^{k(1-\xi),k;\psi}_{a+} x(a) + \left(\mathcal{J}^{\alpha,k;\psi}_{a+} (\tilde{w} + \sigma) \right)(t),$$

where $\tilde{w} \in C^1_{\xi;\psi}(J)$ be a function satisfying the functional equation

$$\tilde{w}(t) = f(t, x(t), \tilde{w}(t)).$$

Hence, for each $t \in J$, we have

$$|x(t) - y(t)| \leq \left(\mathcal{J}^{\alpha,k;\psi}_{a+}|\tilde{w}(s) - w(s)|\right)(t) + \left(\mathcal{J}^{\alpha,k;\psi}_{a+}\sigma\right)(t)$$

$$\leq \epsilon\kappa_v v(t) + \frac{\eta_1}{(1-\eta_2)k\Gamma_k(\alpha)} \int_a^t \frac{\psi'(s)|x(s) - y(s)|dt}{(\psi(t) - \psi(s))^{1-\frac{\alpha}{k}}}.$$

By applying Theorem 2.42, we obtain

$$|x(t) - y(t)| \leq \epsilon\kappa_v v(t) + \int_a^t \sum_{i=1}^{\infty} \frac{\left(\frac{\eta_1}{1-\eta_2}\right)^i}{k\Gamma_k(\alpha i)} \psi'(s) \left[\psi(t) - \psi(s)\right]^{\frac{i\alpha}{k}-1} \epsilon\kappa_v v(s)\,ds,$$

$$\leq \epsilon\kappa_v v(t)\mathbb{E}^{\alpha,k}_k \left[\frac{\eta_1}{1-\eta_2}(\psi(t) - \psi(a))^{\frac{\alpha}{k}}\right]$$

$$\leq \epsilon\kappa_v v(t)\mathbb{E}^{\alpha,k}_k \left[\frac{\eta_1}{1-\eta_2}(\psi(b) - \psi(a))^{\frac{\alpha}{k}}\right].$$

Then for each $t \in J$, we have

$$|x(t) - y(t)| \leq a_{f,v}\epsilon v(t),$$

where

$$a_{f,v} = \kappa_v \mathbb{E}^{\alpha,k}_k \left[\frac{\eta_1}{1-\eta_2}(\psi(b) - \psi(a))^{\frac{\alpha}{k}}\right].$$

Hence, the problem (3.23)–(3.24) is U-H-R stable with respect to v.   □

3.3.3   Examples

With the following examples, we look at particular cases of the problem (3.23)–(3.24).

Example 3.18 Taking $\beta \to 0$, $\alpha = \frac{1}{2}$, $k = 1$, $\psi(t) = t$, $a = 1$, $b = 2$, and $x_0 = 1$, we obtain a particular case of problem (3.23)–(3.24) with the Riemann-Liouville fractional derivative, given by

$$\left({}^H_1\mathcal{D}^{\frac{1}{2},0;\psi}_{1+}x\right)(t) = \left({}^{RL}\mathcal{D}^{\frac{1}{2}}_{1+}x\right)(t) = f\left(t, x(t), \left({}^{RL}\mathcal{D}^{\frac{1}{2}}_{1+}x\right)(t)\right), \quad t \in (1, 2], \quad (3.32)$$

$$\left(\mathcal{J}^{\frac{1}{2},1;\psi}_{1+}x\right)(1^+) = 1, \tag{3.33}$$

where $\bar{J} = [1, 2]$, $\xi = \frac{1}{k}(\beta(k - \alpha) + \alpha) = \frac{1}{2}$, and

$$f(t, x, y) = \frac{\sqrt{t - 1}|sin(t)|(1 + x + y)}{66e^{-t+3}}, \quad t \in \bar{J}, \; x, y \in \mathbb{R}.$$

We have

$$C_{\xi;\psi}(J) = C_{\frac{1}{2};\psi}(J) = \left\{ u : (1, 2] \rightarrow \mathbb{R} : (\sqrt{t - 1})u \in C(\bar{J}, \mathbb{R}) \right\},$$

and

$$C_{\xi;\psi}^1(J) = C_{\frac{1}{2};\psi}^1(J) = \left\{ u \in C_{\frac{1}{2};\psi}(J) : u' \in C_{\frac{1}{2};\psi}(J) \right\}.$$

Since the continuous function $f \in C_{\frac{1}{2};\psi}^1(J)$, then the condition (3.14.1) is satisfied. For each $x, \bar{x}, y, \bar{y} \in \mathbb{R}$ and $t \in \bar{J}$, we have

$$|f(t, x, \bar{x}) - f(t, y, \bar{y})| \leq \frac{\sqrt{t - 1}|sin(t)|}{66e^{-t+3}} (|x - \bar{x}| + |y - \bar{y}|),$$

and so the condition (3.14.2) is satisfied with $\eta_1 = \eta_2 = \dfrac{1}{66e}$. Also, the condition (3.29) of Theorem 3.14 is satisfied. Indeed, we have

$$\mathcal{L} = \frac{\sqrt{\pi}}{66e - 1} \approx 0.01 \; < 1.$$

Then the problem (3.32)–(3.33) has a unique solution in $C_{\frac{1}{2};\psi}^1([1, 2])$.

Now, if we take $v(t) = t - 1$ and $\kappa_v = \dfrac{\sqrt{2}\Gamma(2)}{\Gamma(\frac{5}{2})}$, then for each $t \in J$, we get

$$\left(\mathcal{J}_{1+}^{\frac{1}{2},1;\psi} v \right)(t) \leq \frac{\sqrt{2}\Gamma(2)}{\Gamma(\frac{5}{2})}(t - 1)$$

$$= \kappa_v v(t),$$

which shows that the hypothesis (3.17.1) is satisfied. Consequently, Theorem 3.17 implies that the problem (3.32)–(3.33) is U-H-R stable.

Example 3.19 Taking $\beta \rightarrow 0$, $\alpha = \frac{1}{2}$, $k = 1$, $\psi(t) = \ln t$, $a = 1$, $b = e$, and $x_0 = \pi$, we get a particular case of problem (3.23)–(3.24) using the Hadamard fractional derivative, given by

$$\left({}_1^H \mathcal{D}_{1+}^{\frac{1}{2},0;\psi} x \right)(t) = \left({}^{HD}\mathcal{D}_{1+}^{\frac{1}{2}} x \right)(t) = f\left(t, x(t), \left({}^{HD}\mathcal{D}_{1+}^{\frac{1}{2}} x \right)(t) \right), \quad t \in (1, e], \quad (3.34)$$

$$\left(\mathcal{J}_{1+}^{\frac{1}{2},1;\psi} x \right)(1^+) = \pi, \quad\quad\quad\quad\quad\quad\quad (3.35)$$

where $\bar{J} = [1, e]$, and

$$f(t, x, y) = \frac{e + x + y}{111e^t}, \ t \in \bar{J}, \ x, y \in \mathbb{R}.$$

We have

$$C_{\xi;\psi}(J) = C_{\frac{1}{2};\psi}(J) = \left\{ u : (1, e] \to \mathbb{R} : (\sqrt{\ln t})u \in C(\bar{J}, \mathbb{R}) \right\},$$

and

$$C^1_{\xi;\psi}(J) = C^1_{\frac{1}{2};\psi}(J) = \left\{ u \in C_{\frac{1}{2};\psi}(J) : u' \in C_{\frac{1}{2};\psi}(J) \right\}.$$

Clearly, the function $f \in C^1_{\frac{1}{2};\psi}(J)$. Hence, condition (3.14.1) is satisfied.
For each $x, \bar{x}, y, \bar{y} \in \mathbb{R}$ and $t \in \bar{J}$, we have

$$|f(t, x, \bar{x}) - f(t, y, \bar{y})| \le \frac{1}{111e^t}|x - \bar{x}| + \frac{1}{111e^t}|y - \bar{y}|,$$

and so the condition (3.14.2) is satisfied with $\eta_1 = \eta_2 = \frac{1}{111e}$.
   Also, we have

$$\mathcal{L} = \frac{\sqrt{\pi}}{111e - 1} \approx 0.0058 \ < 1,$$

then the condition (3.29) of Theorem 3.14 is satisfied. Then the problem (3.34)–(3.35) has a unique solution in $C_{\frac{1}{2};\psi}([1, e])$. The problem is also U-H-R stable if we take $v(t) = e^2$ and $\kappa_v = \frac{1}{\Gamma(\frac{3}{2})}$. Indeed, for each $t \in J$, we get

$$\left(\mathcal{J}^{\frac{1}{2}, 1; \psi}_{1+} v \right)(t) \le \frac{e^2}{\Gamma(\frac{3}{2})}$$
$$= \kappa_v v(t).$$

3.4   Existence and Ulam Stability Results for k-Generalized ψ-Hilfer Initial Value Problem in Banach Spaces

This section deals with the initial value problem with nonlinear implicit k-generalized ψ-Hilfer-type fractional differential equation :

$$\left({}^H_k \mathcal{D}^{\alpha, \beta; \psi}_{a+} x \right)(t) = f\left(t, x(t), \left({}^H_k \mathcal{D}^{\alpha, \beta; \psi}_{a+} x \right)(t) \right), \ t \in J, \qquad (3.36)$$

$$\left(\mathcal{J}^{k(1-\xi), k; \psi}_{a+} x \right)(a^+) = x_0, \qquad (3.37)$$

where $^{H}_{k}\mathcal{D}^{\alpha,\beta;\psi}_{a+}$, $\mathcal{J}^{k(1-\xi),k;\psi}_{a+}$ are the k-generalized ψ-Hilfer fractional derivative of order $\alpha \in (0,1)$ and type $\beta \in [0,1]$, and k-generalized ψ-fractional integral of order $k(1-\xi)$, where $\xi = \frac{1}{k}(\beta(k-\alpha)+\alpha)$, $x_0 \in E$, $k > 0$, and $f \in C(\bar{J} \times E \times E, E)$.

3.4.1 Existence Results

We consider the following fractional differential equation:

$$\left(^{H}_{k}\mathcal{D}^{\alpha,\beta;\psi}_{a+} x\right)(t) = w(t), \quad t \in J, \tag{3.38}$$

where $0 < \alpha < 1, 0 \le \beta \le 1$, with the condition

$$\left(\mathcal{J}^{k(1-\xi),k;\psi}_{a+} x\right)(a^{+}) = x_0, \tag{3.39}$$

where $\xi = \dfrac{\beta(k-\alpha)+\alpha}{k}$, $x_0 \in E$, $k > 0$, and where $w \in C(\bar{J}, E)$ satisfies the functional equation:

$$w(t) = f(t, x(t), w(t)).$$

Following the same approach of Theorem 3.12, we have the results that follow.

Theorem 3.20 *If $w(\cdot) \in C^{1}_{\xi;\psi}(J)$, then x satisfies (3.38)–(3.39) if and only if it satisfies*

$$x(t) = \frac{(\psi(t)-\psi(a))^{\xi-1}}{\Gamma_k(k\xi)} x_0 + \left(\mathcal{J}^{\alpha,k;\psi}_{a+} w\right)(t). \tag{3.40}$$

As a consequence of Theorem 3.20, we have the following result.

Lemma 3.21 *Let $\xi = \dfrac{\beta(k-\alpha)+\alpha}{k}$ where $0 < \alpha < 1, 0 \le \beta \le 1$ and $k > 0$, let $f : \bar{J} \times E \times E \to E$ be a continuous function such that $f(\cdot, x(\cdot), y(\cdot)) \in C^{1}_{\xi;\psi}(J)$, for any $x, y \in C_{\xi;\psi}(J)$. Then, x satisfies the problem (3.36)–(3.37) if and only if x is the fixed point of the operator $T : C_{\xi;\psi}(J) \to C_{\xi;\psi}(J)$ defined by*

$$(Tx)(t) = \frac{(\psi(t)-\psi(a))^{\xi-1}}{\Gamma_k(k\xi)} x_0 + \frac{1}{k\Gamma_k(\alpha)} \int_a^t \frac{\psi'(s)\varphi(s)ds}{(\psi(t)-\psi(s))^{1-\frac{\alpha}{k}}}, \tag{3.41}$$

where φ be a function satisfying the functional equation

$$\varphi(t) = f(t, x(t), \varphi(t)).$$

Lemma 3.22 ([75]) *Let $D \subset C_{\xi;\psi}(J)$ be a bounded and equicontinuous set , then*
(i) the function $t \to \mu(D(t))$ is continuous on J, and

$$\mu_{C_{\xi;\psi}}(D) = \sup_{t \in \bar{J}} \mu\left((\psi(t) - \psi(a))^{1-\xi} D(t)\right).$$

(ii) $\mu\left(\int_a^b u(s)ds : u \in D\right) \le \int_a^b \mu(D(s))ds$, *where*

$$D(t) = \{u(t) : t \in D\}, t \in J.$$

Now we can state and demonstrate our existence result for the problem (3.36)–(3.37) by using Mönch's fixed point Theorem.

Theorem 3.23 *Suppose that the assumptions that follow hold.*

(3.23.1) *The function $t \mapsto f(t, x, y)$ is measurable and continuous on J for each $x, y \in E$, the functions $x \mapsto f(t, x, y)$ and $y \mapsto f(t, x, y)$ are continuous on E for $t \in J$ and*

$$f(\cdot, x(\cdot), y(\cdot)) \in C^1_{\xi;\psi}(J), \text{ for any } x, y \in C_{\xi;\psi}(J).$$

(3.23.2) *There exists a continuous function $p : \bar{J} \longrightarrow [0, \infty)$ such that*

$$\|f(t, x, y)\| \le p(t), \text{ for } t \in J \text{ and for each } x, y \in E.$$

(3.23.3) *For each bounded and measurable set $B \subset E$ and for each $t \in J$, we have*

$$\mu\left(f\left(t, B, \left({}^H_k\mathcal{D}^{\alpha,\beta;\psi}_{a+}B\right)\right)\right) \le (\psi(t) - \psi(a))^{1-\xi} p(t)\mu(B)$$

where ${}^H_k\mathcal{D}^{\alpha,\beta;\psi}_{a+}B = \left\{{}^H_k\mathcal{D}^{\alpha,\beta;\psi}_{a+}w : w \in B\right\}$ *and* $p^* = \sup_{t \in \bar{J}} p(t)$.

If

$$\mathcal{L} = \frac{p^*\, (\psi(b) - \psi(a))^{1-\xi+\frac{\alpha}{k}}}{\Gamma_k(\alpha + k)} < 1, \tag{3.42}$$

then the problem (3.36)–(3.37) has at least one solution in $C_{\xi;\psi}(J)$.

Proof The proof will be given in several steps.

Step 1: We show that the operator $\mathcal{T}$ defined in (3.41) transforms the ball $B_R := B(0, R) = \{w \in C_{\xi;\psi}(J) : \|w\|_{C_{\xi;\psi}} \le R\}$ into itself.
For any $x \in C_{\xi;\psi}(J)$, and each $t \in J$ we have

$$\left\| (\psi(t) - \psi(a))^{1-\xi} \, (\mathcal{T}x)(t) \right\| \leq \frac{\|x_0\|}{\Gamma_k(k\xi)} + \frac{(\psi(t) - \psi(a))^{1-\xi}}{k\Gamma_k(\alpha)} \int_a^t \frac{\psi'(s)\|\varphi(s)\|ds}{(\psi(t) - \psi(s))^{1-\frac{\alpha}{k}}}$$

$$\leq \frac{\|x_0\|}{\Gamma_k(k\xi)} + p^* \, (\psi(t) - \psi(a))^{1-\xi} \left(\mathcal{J}_{a+}^{\alpha,k;\psi}(1) \right)(t).$$

By Lemma 2.21, we have

$$\left\| (\psi(t) - \psi(a))^{1-\xi} \, (\mathcal{T}x)(t) \right\| \leq \frac{\|x_0\|}{\Gamma_k(k\xi)} + \frac{p^* \, (\psi(t) - \psi(a))^{1-\xi+\frac{\alpha}{k}}}{\Gamma_k(\alpha + k)}.$$

Hence , for any $x \in C_{\xi;\psi}(J)$, and each $t \in J$ we get

$$\|\mathcal{T}x\|_{C_{\xi;\psi}} \leq \frac{\|x_0\|}{\Gamma_k(k\xi)} + \frac{p^* \, (\psi(b) - \psi(a))^{1-\xi+\frac{\alpha}{k}}}{\Gamma_k(\alpha + k)} := R.$$

Step 2: $\mathcal{T} : B_R \to B_R$ is continuous.
Let $\{x_n\}_{n\in\mathbb{N}}$ be a sequence such that $x_n \longrightarrow x$ in B_R . Then, for each $t \in J$, we have

$$\left\| (\psi(t) - \psi(a))^{1-\xi} \, [(\mathcal{T}x_n)(t) - (\mathcal{T}x)(t)] \right\| \leq \frac{(\psi(t) - \psi(a))^{1-\xi}}{k\Gamma_k(\alpha)}$$

$$\times \int_a^t \frac{\psi'(s)\|\varphi_n(s) - \varphi(s)\|ds}{(\psi(t) - \psi(s))^{1-\frac{\alpha}{k}}},$$

where $\varphi_n, \varphi \in C_{\xi;\psi}(J)$ such that

$$\varphi_n(t) = f(t, x_n(t), \varphi_n(t)),$$
$$\varphi(t) = f(t, x(t), \varphi(t)).$$

Since $x_n \longrightarrow x$ as $n \longrightarrow \infty$ and f is continuous , then by the Lebesgue dominated convergence theorem, we have

$$\|\mathcal{T}x_n - \mathcal{T}x\|_{C_{\xi;\psi}} \longrightarrow 0 \text{ as } n \longrightarrow \infty.$$

Step 3: $\mathcal{T}(B_R)$ is bounded and equicontinuous. Since $\mathcal{T}(B_R) \subset B_R$ and B_R is bounded , then $\mathcal{T}(B_R)$ is bounded.
Next , let $t_1, t_2 \in J$ such that $a < t_1 < t_2 \leq b$ and let $u \in B_R$. Thus, we have

$$\left\| (\psi(t_2) - \psi(a))^{1-\xi} \, (\mathcal{T}x)(t_2) - (\psi(t_1) - \psi(a))^{1-\xi} \, (\mathcal{T}x)(t_1) \right\|$$

$$\leq \left\| \frac{(\psi(t_2) - \psi(a))^{1-\xi}}{k\Gamma_k(\alpha)} \int_a^{t_2} \frac{\psi'(s)\varphi(s)ds}{(\psi(t_2) - \psi(s))^{1-\frac{\alpha}{k}}} \right.$$

$$\left. - \frac{(\psi(t_1) - \psi(a))^{1-\xi}}{k\Gamma_k(\alpha)} \int_a^{t_1} \frac{\psi'(s)\varphi(s)ds}{(\psi(t_1) - \psi(s))^{1-\frac{\alpha}{k}}} \right\|$$

$$\leq \frac{(\psi(t_2) - \psi(a))^{1-\xi}}{k\Gamma_k(\alpha)} \int_{t_1}^{t_2} \frac{\psi'(s)\|\varphi(s)\|ds}{(\psi(t_2) - \psi(s))^{1-\frac{\alpha}{k}}}$$

$$+ \frac{1}{k\Gamma_k(\alpha)} \int_a^{t_1} \left| \frac{(\psi(t_2) - \psi(a))^{1-\xi}}{(\psi(t_2) - \psi(s))^{1-\frac{\alpha}{k}}} - \frac{(\psi(t_1) - \psi(a))^{1-\xi}}{(\psi(t_1) - \psi(s))^{1-\frac{\alpha}{k}}} \right| \psi'(s)\|\varphi(s)\|ds$$

$$\leq p^* (\psi(b) - \psi(a))^{1-\xi} \left(\mathcal{J}_{t_1+}^{\alpha,k;\psi}(1) \right)(t_2)$$

$$+ \frac{p^*}{k\Gamma_k(\alpha)} \int_a^{t_1} \left| \frac{(\psi(t_2) - \psi(a))^{1-\xi}}{(\psi(t_2) - \psi(s))^{1-\frac{\alpha}{k}}} - \frac{(\psi(t_1) - \psi(a))^{1-\xi}}{(\psi(t_1) - \psi(s))^{1-\frac{\alpha}{k}}} \right| \psi'(s)ds.$$

By Lemma 2.21, we have

$$\left\| (\psi(t_2) - \psi(a))^{1-\xi} \, (\mathcal{T}x)(t_2) - (\psi(t_1) - \psi(a))^{1-\xi} \, (\mathcal{T}x)(t_1) \right\|$$

$$\leq \frac{p^* (\psi(b) - \psi(a))^{1-\xi} (\psi(t_2) - \psi(t_1))^{\frac{\alpha}{k}}}{\Gamma_k(\alpha + k)}$$

$$+ \frac{p^*}{k\Gamma_k(\alpha)} \int_a^{t_1} \left| \frac{(\psi(t_2) - \psi(a))^{1-\xi}}{(\psi(t_2) - \psi(s))^{1-\frac{\alpha}{k}}} - \frac{(\psi(t_1) - \psi(a))^{1-\xi}}{(\psi(t_1) - \psi(s))^{1-\frac{\alpha}{k}}} \right| \psi'(s)ds.$$

As $t_1 \longrightarrow t_2$, the right side of the above inequality tends to zero. Hence, $\mathcal{T}(B_R)$ is bounded and equicontinuous.

Step 4: The implication (2.11) of Theorem 2.49 holds.

Now let D be an equicontinuous subset of B_R such that $D \subset \overline{\mathcal{T}(D)} \cup \{0\}$; therefore, the function $t \longrightarrow d(t) = \mu(D(t))$ is continuous on J. By (3.23.3) and the properties of the measure μ, for each $t \in J$, we have

$$(\psi(t) - \psi(a))^{1-\xi} d(t) \leq \mu \left((\psi(t) - \psi(a))^{1-\xi} (\mathcal{T}D)(t) \cup \{0\} \right)$$

$$\leq \mu \left((\psi(t) - \psi(a))^{1-\xi} (\mathcal{T}D)(t) \right)$$

$$\leq \frac{(\psi(b) - \psi(a))^{1-\xi}}{k\Gamma_k(\alpha)}$$

$$\times \int_a^t \frac{(\psi(s) - \psi(a))^{1-\xi}}{(\psi(t) - \psi(s))^{1-\frac{\alpha}{k}}} \psi'(s) p(s) \mu(D(s)) ds$$

$$\leq p^* (\psi(b) - \psi(a))^{1-\xi} \|d\|_{C_{\xi;\psi}} \left(\mathcal{J}_{a+}^{\alpha,k;\psi}(1) \right)(t)$$

$$\leq \frac{p^* (\psi(b) - \psi(a))^{1-\xi+\frac{\alpha}{k}}}{\Gamma_k(\alpha + k)} \|d\|_{C_{\xi;\psi}}.$$

Thus

$$\|d\|_{C_{\xi;\psi}} \leq \mathcal{L}\|d\|_{C_{\xi;\psi}}.$$

From (3.42), we get $\|d\|_{C_{\xi;\psi}} = 0$, that is $d(t) = \mu(D(t)) = 0$, for each $t \in J$, and then $D(t)$ is relatively compact in E. In view of the Ascoli-Arzela Theorem, D is relatively compact in B_R. Applying now Theorem 2.49, we conclude that $\mathcal{T}$ has a fixed point, which is solution of the problem (3.36)–(3.37). □

Our next existence result for the problem (3.36)–(3.37) is based on Darbo's fixed point theorem.

Theorem 3.24 *Assume that the hypothesis (3.23.1)-(3.23.3) and the condition (3.42) hold. Then the problem (3.36)–(3.37) has a solution defined on J.*

Proof Consider the operator $\mathcal{T}$ is defined as in (3.41). We shall show that $\mathcal{T}$ satisfies the assumption of Darbo's fixed point theorem.

We know that $\mathcal{T} : B_R \longrightarrow B_R$ is bounded and continuous and that $\mathcal{T}(B_R)$ is equicontinuous, we need to prove that the operator $\mathcal{T}$ is a $\mathcal{L}$-set contraction.

Let $D \subset B_R$ and $t \in J$. Then we have

$$\mu\left((\psi(t) - \psi(a))^{1-\xi}(\mathcal{T}D)(t)\right) = \mu\left((\psi(t) - \psi(a))^{1-\xi}(\mathcal{T}x)(t) : x \in D\right)$$

$$\leq \left\{\frac{1}{k\Gamma_k(\alpha)}\int_a^t \frac{(\psi(s) - \psi(a))^{1-\xi}}{(\psi(t) - \psi(s))^{1-\frac{\alpha}{k}}}\psi'(s)p(s)\mu(D(s))ds : x \in D\right\}$$

$$\times (\psi(b) - \psi(a))^{1-\xi}$$

$$\leq p^*(\psi(b) - \psi(a))^{1-\xi}\mu_{C_{\xi;\psi}}(D)\left(\mathcal{J}_{a+}^{\alpha,k;\psi}(1)\right)(t)$$

$$\leq \frac{p^*(\psi(b) - \psi(a))^{1-\xi+\frac{\alpha}{k}}}{\Gamma_k(\alpha+k)}\mu_{C_{\xi;\psi}}(D).$$

Therefore,

$$\mu_{C_{\xi;\psi}}(\mathcal{T}D) \leq \frac{p^*(\psi(b) - \psi(a))^{1-\xi+\frac{\alpha}{k}}}{\Gamma_k(\alpha+k)}\mu_{C_{\xi;\psi}}(D).$$

So, by (3.42),the operator $\mathcal{T}$ is a $\mathcal{L}$-set contraction. Consequently, from Theorem 2.48, we conclude that $\mathcal{T}$ has a fixed point $x \in B_R$ which is a solution to our problem (3.36)–(3.37). □

3.4.2   Ulam-Hyers-Rassias Stability

Now, we consider the Ulam stability for problem (3.36)–(3.37). Let $x \in C^1_{\xi;\psi}(J)$, $\epsilon > 0$, and $v : J \longrightarrow [0, \infty)$ be a continuous function. We consider the following inequality:

$$\left\| \left({}^H_k \mathcal{D}^{\alpha,\beta;\psi}_{a+} x \right)(t) - f\left(t, x(t), \left({}^H_k \mathcal{D}^{\alpha,\beta;\psi}_{a+} x \right)(t) \right) \right\| \leq \epsilon v(t), \ t \in J. \qquad (3.43)$$

Definition 3.25 Problem (3.36)–(3.37) is Ulam-Hyers-Rassias (U-H-R) stable with respect to v if there exists a real number $a_{f,v} > 0$ such that for each $\epsilon > 0$ and for each solution $x \in C^1_{\xi;\psi}(J)$ of inequality (3.43) there exists a solution $y \in C^1_{\xi;\psi}(J)$ of (3.36), (3.37) with

$$\|x(t) - y(t)\| \leq \epsilon a_{f,v} v(t), \qquad t \in J.$$

Remark 3.26 A function $x \in C^1_{\xi;\psi}(J)$ is a solution of inequality (3.43) if and only if there exist $\sigma \in C_{\xi;\psi}(J)$ such that

1. $\|\sigma(t)\| \leq \epsilon v(t), t \in J,$
2. $\left({}^H_k \mathcal{D}^{\alpha,\beta;\psi}_{a+} x \right)(t) = f\left(t, x(t), \left({}^H_k \mathcal{D}^{\alpha,\beta;\psi}_{a+} x \right)(t) \right) + \sigma(t), t \in J.$

Theorem 3.27 *Assume that in addition to (3.23.1)–(3.23.3) and (3.42), the following hypothesis hold.*

(3.27.1) *There exist a nondecreasing function $v \in C^1_{\xi;\psi}(J)$ and $\kappa_v > 0$ such that for each $t \in J$, we have*
$$\left(\mathcal{J}^{\alpha,k;\psi}_{a+} v \right)(t) \leq \kappa_v v(t).$$

(3.27.2) *There exists a continuous function $q : \bar{J} \longrightarrow [0, \infty)$ such that for each $t \in J$, we have*
$$p(t) \leq q(t) v(t).$$

Then the problem (3.36)–(3.37) is U-H-R stable.
    Set $q^* = \sup_{t \in \bar{J}} q(t)$.

Proof Let $x \in C^1_{\xi;\psi}(J)$ be a solution of inequality (3.43), and let us assume that y is the unique solution of the problem

$$\begin{cases} \left({}^H_k \mathcal{D}^{\alpha,\beta;\psi}_{a+} y \right)(t) = f\left(t, y(t), \left({}^H_k \mathcal{D}^{\alpha,\beta;\psi}_{a+} y \right)(t) \right) ; \ t \in J, \\ \left(\mathcal{J}^{k(1-\xi),k;\psi}_{a+} y \right)(a^+) = \left(\mathcal{J}^{k(1-\xi),k;\psi}_{a+} x \right)(a^+). \end{cases}$$

By Lemma 3.21, we obtain for each $t \in J$

$$y(t) = \frac{(\psi(t) - \psi(a))^{\xi-1}}{\Gamma_k(k\xi)} \mathcal{J}^{k(1-\xi),k;\psi}_{a+} y(a) + \left(\mathcal{J}^{\alpha,k;\psi}_{a+} w \right)(t),$$

where $w \in C^1_{\xi;\psi}(J)$ be a function satisfying the functional equation

$$w(t) = f(t, y(t), w(t)).$$

Since x is a solution of the inequality (3.43), by Remark 3.26, we have

$$\left({}^H_k \mathcal{D}^{\alpha,\beta;\psi}_{a+} x \right)(t) = f\left(t, x(t), \left({}^H_k \mathcal{D}^{\alpha,\beta;\psi}_{a+} x \right)(t) \right) + \sigma(t), t \in J. \tag{3.44}$$

Clearly, the solution of (3.44) is given by

$$x(t) = \frac{(\psi(t) - \psi(a))^{\xi-1}}{\Gamma_k(k\xi)} \mathcal{J}^{k(1-\xi),k;\psi}_{a+} x(a) + \left(\mathcal{J}^{\alpha,k;\psi}_{a+} (\tilde{w} + \sigma) \right)(t),$$

where $\tilde{w} \in C^1_{\xi;\psi}(J)$ be a function satisfing the functional equation

$$\tilde{w}(t) = f(t, x(t), \tilde{w}(t)).$$

Hence, for each $t \in J$, we have

$$\|x(t) - y(t)\| \leq \left(\mathcal{J}^{\alpha,k;\psi}_{a+} \|\tilde{w}(s) - w(s)\| \right)(t) + \left(\mathcal{J}^{\alpha,k;\psi}_{a+} \|\sigma(s)\| \right)(t)$$
$$\leq \epsilon \kappa_v v(t) + \frac{1}{k\Gamma_k(\alpha)} \int_a^t \frac{\psi'(s) 2q(s) v(s) ds}{(\psi(t) - \psi(s))^{1-\frac{\alpha}{k}}}$$
$$\leq \epsilon \kappa_v v(t) + 2q^* \left(\mathcal{J}^{\alpha,k;\psi}_{a+} v \right)(t)$$
$$\leq (\epsilon + 2q^*) \kappa_v v(t).$$

Then for each $t \in J$, we have

$$\|x(t) - y(t)\| \leq a_{f,v} \epsilon v(t),$$

where

$$a_{f,v} = \kappa_v \left(1 + \frac{2q^*}{\epsilon} \right).$$

Hence, the problem (3.36)–(3.37) is U-H-R stable.   □

3.4.3   Examples

Let

$$E = l^1 = \left\{ u = (u_1, u_2, \ldots, u_n, \ldots), \sum_{n=1}^{\infty} |u_n| < \infty \right\}$$

be the Banach space with the norm

$$\|u\| = \sum_{n=1}^{\infty} |u_n|.$$

Example 3.28 Taking $\beta \to 0$, $\alpha = \frac{1}{2}$, $k = 1$, $\psi(t) = 2\sqrt{t}$, $a = 1$, $b = 3$, and $x_0 \in E$, we get a problem of generalized Hilfer fractional differential equation of the form

$$\left({}^H_1 D^{\frac{1}{2},0;\psi}_{1^+} x_n \right)(t) = \left(\tfrac{1}{2} D^{\frac{1}{2},0}_{1^+} x_n \right)(t) = f_n \left(t, x_n(t), \left(\tfrac{1}{2} D^{\frac{1}{2},0}_{1^+} x_n \right)(t) \right), t \in (1, 3] \tag{3.45}$$

$$\left(\mathcal{J}^{\frac{1}{2},1;\psi}_{1^+} x \right)(1^+) = x_0, \tag{3.46}$$

where

$$f_n \left(t, x_n(t), \left(\tfrac{1}{2} D^{\frac{1}{2},0}_{1^+} x_n \right)(t) \right) = \frac{(2t^2 + e^{-2}) |x_n(t)|}{177 e^{-t+3} \left(1 + \|x(t)\| + \left\| \left(\tfrac{1}{2} D^{\frac{1}{2},0}_{1^+} x_n \right)(t) \right\| \right)},$$

for $t \in (1, 3]$ with

$$f = (f_1, f_2, \ldots, f_n, \ldots) \text{ and } x = (x_1, x_2, \ldots, x_n, \ldots).$$

Set

$$f(t, x, y) = \frac{(2t^2 + e^{-2}) \|x\|}{177 e^{-t+3} (1 + \|x\| + \|y\|)}, \quad t \in (1, 3], \ x, y \in E.$$

We have

$$C_{\xi;\psi}(J) = C_{\frac{1}{2};\psi}(J) = \left\{ x : (1, 3] \to E : \sqrt{2} \left(\sqrt{t} - 1 \right)^{\frac{1}{2}} x \in C(\bar{J}, E) \right\},$$

and

$$C^1_{\xi,1,\psi}(J) = C^1_{\frac{1}{2};\psi}(J) = \left\{ x \in C_{\frac{1}{2};\psi}(J) : x' \in C_{\frac{1}{2};\psi}(J) \right\}.$$

It is clear that the function f satisfies the hypothesis (3.23.1). Also, the hypothesis (3.23.2) is satisfied with

$$p(t) = \frac{2t^2 + e^{-2}}{177 e^{-t+3}}, t \in (1, 3],$$

and

$$p^* = \frac{18 + e^{-2}}{177}.$$

The conditions of Theorem 3.23 are satisfied. Indeed, we have

$$\mathcal{L} = \frac{p^* \left(\psi(b) - \psi(a)\right)^{1-\xi+\frac{\alpha}{k}}}{\Gamma_k(\alpha + k)}$$

$$= \frac{2(18 + e^{-2})\left(2\sqrt{3} - 2\right)}{177\sqrt{\pi}}$$

$$\approx 0.169269 < 1.$$

Hence, the problem (3.45)–(3.46) has at least one solution defined on $(1, 3]$.
Let
$v(t) = e$, and $\kappa_v = \dfrac{4}{\sqrt{\pi}}$. Then, for each $t \in (1, 3]$, we get

$$\left(\mathcal{J}_{1+}^{\frac{1}{2},1;\psi} v\right)(t) \leq \frac{4e}{\sqrt{\pi}}$$

$$= \kappa_v v(t).$$

Let the function $q : [1, 3] \to [0, \infty)$ be defined by

$$q(t) = \frac{2t^2 + e^{-2}}{177e^{-t+4}},$$

then, for each $t \in (1, 3]$, we have

$$p(t) = q(t)v(t).$$

Then, hypothesis (3.27.1) and (3.27.2) are satisfied; consequently, Theorem 3.27 implies that problem (3.45)–(3.46) is Ulam-Hyers-Rassias stable.

Example 3.29 Taking $\beta \to 1, \alpha = \frac{1}{2}, k = 1, \psi(t) = t, a = 1, b = 2$, and $x_0 \in E$, we get a particular case of problem (3.36)–(3.37), which is a problem of Caputo Fractional differential equation of the form

$$\left({}^H D_{1+}^{\frac{1}{2},1;\psi} x_n\right)(t) = \left({}^C D_{1+}^{\frac{1}{2}} x_n\right)(t) = f_n\left(t, x_n(t), \left({}^C D_{1+}^{\frac{1}{2}} x_n\right)(t)\right), t \in (1, 2] \quad (3.47)$$

$$\left(\mathcal{J}_{1+}^{0,1;\psi} x\right)(1^+) = x_0, \quad (3.48)$$

where

$$f_n\left(t, x_n(t), \left({}^C D_{1+}^{\frac{1}{2}} x_n\right)(t)\right) = \frac{e^{t-2}|x_n(t)|}{33\left(1 + \|x(t)\| + \left\|\left({}^C D_{1+}^{\frac{1}{2}} x_n\right)(t)\right\|\right)}, \quad t \in (1, 2].$$

With

$$f = (f_1, f_2, \ldots, f_n, \ldots) \text{ and } x = (x_1, x_2, \ldots, x_n, \ldots).$$

Set

$$f(t, x, y) = \frac{e^{t-2}\|x\|}{33(1 + \|x\| + \|y\|)}, \quad t \in (1, 2], \ x, y \in E.$$

We have

$$C_{\xi;\psi}(J) = C_{1;\psi}(J) = \left\{x : (1, 2] \to E : x \in C(\bar{J}, E)\right\},$$

and

$$C_{\xi,1,\psi}^1(J) = C_{1;\psi}^1(J) = \left\{x \in C_{1;\psi}(J) : x' \in C_{1;\psi}(J)\right\}.$$

Hence, the function f satisfies the hypothesis (3.23.1). Also, the hypothesis (3.23.2) is satisfied with

$$p(t) = \frac{e^{t-2}}{33}, t \in (1, 2],$$

and

$$p^* = \frac{1}{33}.$$

Since

$$\mathcal{L} = \frac{p^* (\psi(b) - \psi(a))^{1-\xi+\frac{\alpha}{k}}}{\Gamma_k(\alpha + k)}$$

$$= \frac{2}{33\sqrt{\pi}}$$

$$\approx 0.03419 < 1,$$

then the condition of Theorem 3.24 is satisfied. Hence, the problem (3.47)–(3.48) has at least one solution defined on (1, 2]. Also, same as the last example, we can easily choose a function v that satisfies the hypothesis (3.27.1) and (3.27.2), which gives us the Ulam-Hyers-Rassias stability of our problem (3.47)–(3.48).

3.5   Existence and k-Mittag-Leffler-Ulam-Hyers Stability Results of k-Generalized ψ-Hilfer Boundary Valued Problem

In this section, we consider the boundary valued problem with nonlinear implicit k-generalized ψ-Hilfer-type fractional differential equation:

$$\left({}^{H}_{k}\mathcal{D}^{\alpha,\beta;\psi}_{a+} x \right)(t) = f\left(t, x(t), \left({}^{H}_{k}\mathcal{D}^{\alpha,\beta;\psi}_{a+} x \right)(t) \right), \quad t \in J, \tag{3.49}$$

$$c_1 \left(\mathcal{J}^{k(1-\xi),k;\psi}_{a+} x \right)(a^+) + c_2 \left(\mathcal{J}^{k(1-\xi),k;\psi}_{a+} x \right)(b) = c_3, \tag{3.50}$$

where ${}^{H}_{k}\mathcal{D}^{\alpha,\beta;\psi}_{a+}, \mathcal{J}^{k(1-\xi),k;\psi}_{a+}$ are the k-generalized ψ-Hilfer fractional derivative of order $\alpha \in (0, k)$ and type $\beta \in [0, 1]$ defined in Section 2, and k-generalized ψ-fractional integral of order $k(1 - \xi)$ defined in [113], respectively, where $\xi = \frac{1}{k}(\beta(k - \alpha) + \alpha)$, $k > 0$, $f \in C(\bar{J} \times \mathbb{R}^2, \mathbb{R})$, and $c_1, c_2, c_3 \in \mathbb{R}$ such that $c_1 + c_2 \neq 0$.

3.5.1   Existence Results

We consider the following fractional differential equation:

$$\left({}^{H}_{k}\mathcal{D}^{\alpha,\beta;\psi}_{a+} x \right)(t) = \varpi(t), \quad t \in J, \tag{3.51}$$

where $0 < \alpha < k, 0 \leq \beta \leq 1$, with the condition

$$c_1 \left(\mathcal{J}^{k(1-\xi),k;\psi}_{a+} x \right)(a^+) + c_2 \left(\mathcal{J}^{k(1-\xi),k;\psi}_{a+} x \right)(b) = c_3, \tag{3.52}$$

where $\xi = \dfrac{\beta(k - \alpha) + \alpha}{k}$, $k > 0$, $c_1, c_2, c_3 \in \mathbb{R}$ such that $c_1 + c_2 \neq 0$ and where $\varpi \in C(\bar{J}, \mathbb{R})$ satisfies the functional equation:

$$\varpi(t) = f(t, x(t), \varpi(t)).$$

The following theorem shows that the problem (3.51)–(3.52) has a unique solution.

Theorem 3.30 *If* $w(\cdot) \in C^1_{\xi;\psi}(J)$, *then* x *satisfies (3.51)–(3.52) if and only if it satisfies*

$$x(t) = \frac{(\psi(t) - \psi(a))^{\xi-1}}{(c_1 + c_2)\Gamma_k(k\xi)} \left[c_3 - c_2 \left(\mathcal{J}^{k(1-\xi)+\alpha,k;\psi}_{a+} \varpi \right)(b) \right] + \left(\mathcal{J}^{\alpha,k;\psi}_{a+} \varpi \right)(t). \tag{3.53}$$

Proof Assume $x \in C^1_{\xi;\psi}(J)$ satisfies Eqs. (3.51) and (3.52), by applying the fractional integral operator $\mathcal{J}^{\alpha,k;\psi}_{a+}(\cdot)$ on both sides of the fractional equation (3.51), we have

$$\left(\mathcal{J}_{a+}^{\alpha,k;\psi}\,{}^H_k\mathcal{D}_{a+}^{\alpha,\beta;\psi}x\right)(t) = \left(\mathcal{J}_{a+}^{\alpha,k;\psi}\varpi\right)(t),$$

using Theorem 2.35, we get

$$x(t) = \frac{(\psi(t) - \psi(a))^{\xi-1}}{\Gamma_k(k\xi)}\mathcal{J}_{a+}^{k(1-\xi),k;\psi}x(a) + \left(\mathcal{J}_{a+}^{\alpha,k;\psi}\varpi\right)(t). \tag{3.54}$$

Applying $\mathcal{J}_{a+}^{k(1-\xi),k;\psi}(\cdot)$ on both sides of (3.54), using Lemmas 2.18, 2.21, and taking $t = b$, we have

$$\left(\mathcal{J}_{a+}^{k(1-\xi),k;\psi}x\right)(b) = \mathcal{J}_{a+}^{k(1-\xi),k;\psi}x(a) + \left(\mathcal{J}_{a+}^{k(1-\xi)+\alpha,k;\psi}\varpi\right)(b). \tag{3.55}$$

Multiplying both sides of (3.55) by c_2, we get

$$c_2\left(\mathcal{J}_{a+}^{k(1-\xi),k;\psi}x\right)(b) = c_2\mathcal{J}_{a+}^{k(1-\xi),k;\psi}x(a) + c_2\left(\mathcal{J}_{a+}^{k(1-\xi)+\alpha,k;\psi}\varpi\right)(b).$$

Using condition (3.52), we obtain

$$c_2\left(\mathcal{J}_{a+}^{k(1-\xi),k;\psi}x\right)(b) = c_3 - c_1\left(\mathcal{J}_{a+}^{k(1-\xi),k;\psi}x\right)(a^+).$$

Thus

$$c_3 - c_1\left(\mathcal{J}_{a+}^{k(1-\xi),k;\psi}x\right)(a^+) = c_2\mathcal{J}_{a+}^{k(1-\xi),k;\psi}x(a) + c_2\left(\mathcal{J}_{a+}^{k(1-\xi)+\alpha,k;\psi}\varpi\right)(b).$$

Then

$$\left(\mathcal{J}_{a+}^{k(1-\xi),k;\psi}x\right)(a^+) = \frac{c_3}{c_1+c_2} - \frac{c_2}{c_1+c_2}\left(\mathcal{J}_{a+}^{k(1-\xi)+\alpha,k;\psi}\varpi\right)(b). \tag{3.56}$$

Substituting (3.56) into (3.54), we obtain (3.53).

Let us now prove that if x satisfies Eq. (3.53), then it satisfies Eqs. (3.51) and (3.52). Applying the fractional derivative operator ${}^H_k\mathcal{D}_{a+}^{\alpha,\beta;\psi}(\cdot)$ on both sides of the fractional equation (3.53), then we get

$$\left({}^H_k\mathcal{D}_{a+}^{\alpha,\beta;\psi}x\right)(t) = {}^H_k\mathcal{D}_{a+}^{\alpha,\beta;\psi}\left(\frac{(\psi(t)-\psi(a))^{\xi-1}}{(c_1+c_2)\Gamma_k(k\xi)}\left[c_3 - c_2\left(\mathcal{J}_{a+}^{k(1-\xi)+\alpha,k;\psi}\varpi\right)(b)\right]\right)$$
$$+ \left({}^H_k\mathcal{D}_{a+}^{\alpha,\beta;\psi}\mathcal{J}_{a+}^{\alpha,k;\psi}\varpi\right)(t).$$

Using the Lemmas 2.31 and 2.29, we obtain Eq. (3.51). Now we apply the operator $\mathcal{J}_{a+}^{k(1-\xi),k;\psi}(\cdot)$ on Eq. (3.53), to have

$$\left(\mathcal{J}_{a+}^{k(1-\xi),k;\psi}x\right)(t) = \left[\frac{c_3 - c_2\left(\mathcal{J}_{a+}^{k(1-\xi)+\alpha,k;\psi}\varpi\right)(b)}{(c_1+c_2)\Gamma_k(k\xi)}\right]\mathcal{J}_{a+}^{k(1-\xi),k;\psi}(\psi(t)-\psi(a))^{\xi-1}$$

$$+\left(\mathcal{J}_{a+}^{k(1-\xi),k;\psi}\mathcal{J}_{a+}^{\alpha,k;\psi}\varpi\right)(t).$$

Now, using Lemmas 2.18 and 2.21, we get

$$\left(\mathcal{J}_{a+}^{k(1-\xi),k;\psi}x\right)(t) = \frac{c_3}{c_1+c_2} - \frac{c_2}{c_1+c_2}\left(\mathcal{J}_{a+}^{k(1-\xi)+\alpha,k;\psi}\varpi\right)(b)$$

$$+\left(\mathcal{J}_{a+}^{k(1-\xi)+\alpha,k;\psi}\varpi\right)(t). \tag{3.57}$$

Using Theorem 2.26 with $t \to a$, we obtain

$$\left(\mathcal{J}_{a+}^{k(1-\xi),k;\psi}x\right)(a^+) = \frac{c_3}{c_1+c_2} - \frac{c_2}{c_1+c_2}\left(\mathcal{J}_{a+}^{k(1-\xi)+\alpha,k;\psi}\varpi\right)(b). \tag{3.58}$$

Now, taking $t = b$ in (3.57), to get

$$\left(\mathcal{J}_{a+}^{k(1-\xi),k;\psi}x\right)(b) = \frac{c_3}{c_1+c_2} - \frac{c_2}{c_1+c_2}\left(\mathcal{J}_{a+}^{k(1-\xi)+\alpha,k;\psi}\varpi\right)(b)$$

$$+\left(\mathcal{J}_{a+}^{k(1-\xi)+\alpha,k;\psi}\varpi\right)(b). \tag{3.59}$$

From (3.58) and (3.59), we obtain (3.52). This completes the proof.   □

As a consequence of Theorem 3.30, we have the following result.

Lemma 3.31 *Let* $\xi = \dfrac{\beta(k-\alpha)+\alpha}{k}$ *where* $0 < \alpha < k$ *and* $0 \le \beta \le 1$, *let* $f : \bar{J} \times \mathbb{R} \times$ $\mathbb{R} \to \mathbb{R}$ *be a continuous function such that* $f(\cdot, x(\cdot), y(\cdot)) \in C_{\xi;\psi}^1(J)$, *for any* $x, y \in$ $C_{\xi;\psi}(J)$. *If* $x \in C_{\xi;\psi}^1(J)$, *then* x *satisfies the problem* (3.49)–(3.50) *if and only if* x *is the fixed point of the operator* $T : C_{\xi;\psi}(J) \to C_{\xi;\psi}(J)$ *defined by*

$$(Tx)(t) = \frac{(\psi(t)-\psi(a))^{\xi-1}}{(c_1+c_2)\Gamma_k(k\xi)}\left[c_3 - c_2\left(\mathcal{J}_{a+}^{k(1-\xi)+\alpha,k;\psi}\varphi\right)(b)\right] + \left(\mathcal{J}_{a+}^{\alpha,k;\psi}\varphi\right)(t), \tag{3.60}$$

where φ *be a function satisfying the functional equation*

$$\varphi(t) = f(t, x(t), \varphi(t)).$$

We are now in a position to state and prove our existence result for the problem (3.49)–(3.50) based on Banach's fixed point theorem [72].

Theorem 3.32 *Assume that the requirements that follow are met.*

(3.32.1) *The function $f : \bar{J} \times \mathbb{R} \times \mathbb{R} \to \mathbb{R}$ is continuous and*

$$f(\cdot, x(\cdot), y(\cdot)) \in C^1_{\xi;\psi}(J), \text{ for any } x, y \in C_{\xi;\psi}(J).$$

(3.32.2) *There exist constants $\eta_1 > 0$ and $0 < \eta_2 < 1$ such that*

$$|f(t, x, y) - f(t, \bar{x}, \bar{y})| \le \eta_1 |x - \bar{x}| + \eta_2 |y - \bar{y}|$$

for any $x, y, \bar{x}, \bar{y} \in \mathbb{R}$ and $t \in \bar{J}$.

If

$$\mathcal{L} = \frac{\eta_1 \left(\psi(b) - \psi(a) \right)^{\frac{\alpha}{k}}}{1 - \eta_2} \left[\frac{|c_2|}{|c_1 + c_2| \Gamma_k(k + \alpha)} + \frac{\Gamma_k(k\xi)}{\Gamma_k(\alpha + k\xi)} \right] < 1, \qquad (3.61)$$

then the problem (3.49)–(3.50) has a unique solution in $C_{\xi;\psi}(J)$.

Proof We show that the operator $\mathcal{T}$ defined in (3.60) has a unique fixed point in $C_{\xi;\psi}(J)$. Let $x, y \in C_{\xi;\psi}(J)$. Then, for $t \in J$ we have

$$\begin{aligned}
|\mathcal{T}x(t) - \mathcal{T}y(t)| &\le \frac{|c_2|(\psi(t) - \psi(a))^{\xi-1}}{|c_1 + c_2| \Gamma_k(k\xi)} \left(\mathcal{J}_{a+}^{k(1-\xi)+\alpha, k;\psi} |\varphi_1(s) - \varphi_2(s)| \right)(b) \\
&\quad + \left(\mathcal{J}_{a+}^{\alpha, k;\psi} |\varphi_1(s) - \varphi_2(s)| \right)(t),
\end{aligned}$$

where φ_1 and φ_1 be functions satisfying the functional equations

$$\begin{aligned}
\varphi_1(t) &= f(t, x(t), \varphi_1(t)), \\
\varphi_2(t) &= f(t, y(t), \varphi_2(t)).
\end{aligned}$$

By (3.32.2), we have

$$\begin{aligned}
|\varphi_1(t) - \varphi_2(t)| &= |f(t, x(t), \varphi_1(t)) - f(t, y(t), \varphi_2(t))| \\
&\le \eta_1 |x(t) - y(t)| + \eta_2 |\varphi_1(t) - \varphi_2(t)|.
\end{aligned}$$

Then,

$$|\varphi_1(t) - \varphi_2(t)| \le \frac{\eta_1}{1 - \eta_2} |x(t) - y(t)|.$$

Therefore, for each $t \in J$

$$|\mathcal{T}x(t) - \mathcal{T}y(t)| \leq \frac{\eta_1 |c_2| (\psi(t) - \psi(a))^{\xi-1} \left(\mathcal{J}_{a+}^{k(1-\xi)+\alpha,k;\psi} |x(s) - y(s)| \right)(b)}{(1-\eta_2)|c_1 + c_2| \Gamma_k(k\xi)}$$

$$+ \frac{\eta_1}{(1-\eta_2)} \left(\mathcal{J}_{a+}^{\alpha,k;\psi} |x(s) - y(s)| \right)(t)$$

$$\leq \left[\frac{|c_2| (\psi(t) - \psi(a))^{\xi-1} \left(\mathcal{J}_{a+}^{k(1-\xi)+\alpha,k;\psi} (\psi(s) - \psi(a))^{\xi-1} \right)(b)}{|c_1 + c_2| \Gamma_k(k\xi)} \right.$$

$$\left. + \left(\mathcal{J}_{a+}^{\alpha,k;\psi} (\psi(s) - \psi(a))^{\xi-1} \right)(t) \right] \frac{\eta_1 \|x - y\|_{C_{\xi;\psi}}}{1 - \eta_2}.$$

By Lemma 2.21, we have

$$|\mathcal{T}x(t) - \mathcal{T}y(t)| \leq \left[\frac{\eta_1 |c_2| (\psi(t) - \psi(a))^{\xi-1} (\psi(b) - \psi(a))^{\frac{\alpha}{k}}}{(1-\eta_2)|c_1 + c_2| \Gamma_k(k+\alpha)} \right.$$

$$\left. + \frac{\eta_1 \Gamma_k(k\xi)}{\Gamma_k(\alpha + k\xi)(1 - \eta_2)} (\psi(t) - \psi(a))^{\frac{\alpha+k\xi}{k}-1} \right] \|x - y\|_{C_{\xi;\psi}}.$$

Hence

$$\left| (\psi(t) - \psi(a))^{1-\xi} (\mathcal{T}x(t) - \mathcal{T}y(t)) \right| \leq \left[\frac{\eta_1 |c_2| (\psi(b) - \psi(a))^{\frac{\alpha}{k}}}{(1-\eta_2)|c_1 + c_2| \Gamma_k(k+\alpha)} \right.$$

$$\left. + \frac{\eta_1 \Gamma_k(k\xi) (\psi(t) - \psi(a))^{\frac{\alpha}{k}}}{\Gamma_k(\alpha + k\xi)(1 - \eta_2)} \right] \|x - y\|_{C_{\xi;\psi}},$$

which implies that

$$\|\mathcal{T}x - \mathcal{T}y\|_{C_{\xi;\psi}} \leq \left[\frac{|c_2|}{|c_1 + c_2| \Gamma_k(k+\alpha)} + \frac{\Gamma_k(k\xi)}{\Gamma_k(\alpha + k\xi)} \right]$$

$$\times \frac{\eta_1 (\psi(b) - \psi(a))^{\frac{\alpha}{k}} \|x - y\|_{C_{\xi;\psi}}}{1 - \eta_2}.$$

By (3.61), the operator $\mathcal{T}$ is a contraction. Hence, by Banach's contraction principle, $\mathcal{T}$ has a unique fixed point $x \in C_{\xi;\psi}(J)$, which is a solution to our problem (3.49)–(3.50).   □

3.5.2 k-Mittag-Leffler-Ulam-Hyers Stability

In this section, we consider the k-Mittag-Leffler-Ulam-Hyers stability for our problem (3.49)–(3.50). Let $x \in C_{\xi;\psi}^1(J)$, $t \in J$, and $\epsilon > 0$. We consider the following inequality:

$$\left| \left({}_k^H \mathcal{D}_{a+}^{\alpha,\beta;\psi} x \right)(t) - f\left(t, x(t), \left({}_k^H \mathcal{D}_{a+}^{\alpha,\beta;\psi} x \right)(t) \right) \right| \leq \epsilon \mathbb{E}_k^\alpha \left((\psi(t) - \psi(a))^{\frac{\alpha}{k}} \right). \quad (3.62)$$

In [94], Liu *et al.* introduced the concept of Ulam-Hyers-Mittag-Leffler; by substituting the Mittag-Leffler function of their definitions with the more refined k-Mittag-Leffler function, we give the following definitions.

Definition 3.33 Problems (3.49)–(3.50) is k-Mittag-Leffler-Ulam-Hyers stable with respect to $\mathbb{E}_k^\alpha\left((\psi(t) - \psi(a))^{\frac{\alpha}{k}}\right)$ if there exists a real number $a_{\mathbb{E}_k^\alpha} > 0$ such that for each $\epsilon > 0$ and for each solution $x \in C_{\xi;\psi}^1(J)$ of inequality (3.62) there exists a solution $y \in C_{\xi;\psi}^1(J)$ of (3.49)–(3.50) with

$$|x(t) - y(t)| \le a_{\mathbb{E}_k^\alpha} \epsilon \mathbb{E}_k^\alpha\left((\psi(t) - \psi(a))^{\frac{\alpha}{k}}\right), \qquad t \in J.$$

Definition 3.34 Problem (3.49)–(3.50) is generalized k-Mittag-Leffler-Ulam-Hyers stable with respect to $\mathbb{E}_k^\alpha\left((\psi(t) - \psi(a))^{\frac{\alpha}{k}}\right)$ if there exists $v : C(\mathbb{R}^+, \mathbb{R}^+)$ with $v(0) = 0$ such that for each $\epsilon > 0$ and for each solution $x \in C_{\xi;\psi}^1(J)$ of inequality (3.62) there exists a solution $y \in C_{\xi;\psi}^1(J)$ of (3.49)–(3.50) with

$$|x(t) - y(t)| \le v(\epsilon)\mathbb{E}_k^\alpha\left((\psi(t) - \psi(a))^{\frac{\alpha}{k}}\right), \qquad t \in J.$$

Remark 3.35 It is clear that Definition 3.33 $\Longrightarrow$ Definition 3.34.

Remark 3.36 A function $x \in C_{\xi;\psi}^1(J)$ is a solution of inequality (3.62) if and only if there exist $\sigma \in C_{\xi;\psi}(J)$ such that

1. $|\sigma(t)| \le \epsilon \mathbb{E}_k^\alpha\left((\psi(t) - \psi(a))^{\frac{\alpha}{k}}\right), t \in J,$
2. $\left(^H_k D_{a+}^{\alpha,\beta;\psi} x\right)(t) = f\left(t, x(t), \left(^H_k D_{a+}^{\alpha,\beta;\psi} x\right)(t)\right) + \sigma(t), t \in J.$

Theorem 3.37 *Assume that the hypothesis (3.32.1), (3.32.2), and the condition (3.61) hold. Then the problem (3.49)–(3.50) is k-Mittag-Leffler-Ulam-Hyers stable with respect to $\mathbb{E}_k^\alpha\left((\psi(t) - \psi(a))^{\frac{\alpha}{k}}\right)$ and consequently generalized k-Mittag-Leffler-Ulam-Hyers stable.*

Proof Let $x \in C_{\xi;\psi}^1(J)$ be a solution of inequality (3.62), and let us assume that y is the unique solution of the problem

$$\begin{cases} \left(^H_k D_{a+}^{\alpha,\beta;\psi} y\right)(t) = f\left(t, y(t), \left(^H_k D_{a+}^{\alpha,\beta;\psi} y\right)(t)\right); \ t \in J, \\ \left(\mathcal{J}_{a+}^{k(1-\xi),k;\psi} y\right)(a^+) = \left(\mathcal{J}_{a+}^{k(1-\xi),k;\psi} x\right)(a^+). \end{cases}$$

By Lemma 3.31, we obtain for each $t \in J$

$$y(t) = \frac{(\psi(t) - \psi(a))^{\xi-1}}{\Gamma_k(k\xi)} \mathcal{J}_{a+}^{k(1-\xi),k;\psi} y(a) + \left(\mathcal{J}_{a+}^{\alpha,k;\psi} w\right)(t),$$

where $w \in C_{\xi;\psi}^1(J)$ be a function satisfying the functional equation

$$w(t) = f(t, y(t), w(t)).$$

Since x is a solution of the inequality (3.62), by Remark 3.36, we have

$$\left({}_k^H \mathcal{D}_{a+}^{\alpha,\beta;\psi} x\right)(t) = f\left(t, x(t), \left({}_k^H \mathcal{D}_{a+}^{\alpha,\beta;\psi} x\right)(t)\right) + \sigma(t), t \in J. \qquad (3.63)$$

Clearly, the solution of (3.63) is given by

$$x(t) = \frac{(\psi(t) - \psi(a))^{\xi-1}}{\Gamma_k(k\xi)} \mathcal{J}_{a+}^{k(1-\xi),k;\psi} x(a) + \left(\mathcal{J}_{a+}^{\alpha,k;\psi} (\tilde{w} + \sigma)\right)(t),$$

where $\tilde{w} \in C_{\xi;\psi}^1(J)$ be a function satisfying the functional equation

$$\tilde{w}(t) = f(t, x(t), \tilde{w}(t)).$$

Hence, for each $t \in J$, we have

$$|x(t) - y(t)|$$
$$\leq \left(\mathcal{J}_{a+}^{\alpha,k;\psi} |\tilde{w}(s) - w(s)|\right)(t) + \left(\mathcal{J}_{a+}^{\alpha,k;\psi} \sigma\right)(t)$$

$$\leq \epsilon \mathcal{J}_{a+}^{\alpha,k;\psi} \mathbb{E}_k^\alpha \left((\psi(t) - \psi(a))^{\frac{\alpha}{k}}\right) + \frac{\eta_1}{(1 - \eta_2)k\Gamma_k(\alpha)} \int_a^t \frac{\psi'(s)|x(s) - y(s)|dt}{(\psi(t) - \psi(s))^{1-\frac{\alpha}{k}}}.$$

Using Lemma 2.39, we get

$$|x(t) - y(t)| \leq \epsilon \mathbb{E}_k^\alpha \left((\psi(t) - \psi(a))^{\frac{\alpha}{k}}\right)$$

$$+ \frac{\eta_1}{(1 - \eta_2)k\Gamma_k(\alpha)} \int_a^t \frac{\psi'(s)|x(s) - y(s)|dt}{(\psi(t) - \psi(s))^{1-\frac{\alpha}{k}}}.$$

By applying Theorem 2.42, we obtain

$|x(t) - y(t)|$

$$\leq \epsilon \mathbb{E}_k^\alpha \left((\psi(t) - \psi(a))^{\frac{\alpha}{k}} \right) + \int_a^t \sum_{i=1}^\infty \frac{\left(\frac{\eta_1}{1-\eta_2} \right)^i \psi'(s) \, \epsilon \mathbb{E}_k^\alpha \left((\psi(s) - \psi(a))^{\frac{\alpha}{k}} \right) ds}{k \Gamma_k(\alpha i) \, [\psi(t) - \psi(s)]^{1 - \frac{i\alpha}{k}}}$$

$$\leq \epsilon \mathbb{E}_k^\alpha \left((\psi(t) - \psi(a))^{\frac{\alpha}{k}} \right) \mathbb{E}_k^\alpha \left[\frac{\eta_1}{1 - \eta_2} (\psi(t) - \psi(a))^{\frac{\alpha}{k}} \right]$$

$$\leq \epsilon \mathbb{E}_k^\alpha \left((\psi(t) - \psi(a))^{\frac{\alpha}{k}} \right) \mathbb{E}_k^\alpha \left[\frac{\eta_1}{1 - \eta_2} (\psi(b) - \psi(a))^{\frac{\alpha}{k}} \right].$$

Then for each $t \in J$, we have

$$|x(t) - y(t)| \leq a_{\mathbb{E}_k^\alpha} \epsilon \mathbb{E}_k^\alpha \left((\psi(t) - \psi(a))^{\frac{\alpha}{k}} \right),$$

where

$$a_{\mathbb{E}_k^\alpha} = \mathbb{E}_k^\alpha \left[\frac{\eta_1}{1 - \eta_2} (\psi(b) - \psi(a))^{\frac{\alpha}{k}} \right].$$

Hence, the problem (3.49)–(3.50) is k-Mittag-Leffler-Ulam-Hyers stable with respect to $\mathbb{E}_k^\alpha \left((\psi(t) - \psi(a))^{\frac{\alpha}{k}} \right)$. If we set $v(\epsilon) = a_{\mathbb{E}_k^\alpha} \epsilon$, then the problem (3.49)–(3.50) is also generalized k-Mittag-Leffler-Ulam-Hyers stable. □

3.5.3   Examples

In this section, we look at particular cases of our problem (3.49)–(3.50), with $\bar{J} = [1, 2]$ and

$$f(t, x, y) = \frac{1 + x + y}{107 e^{-t+2}}, \quad t \in \bar{J}, \; x, y \in \mathbb{R}.$$

$$C_{\xi; \psi}(J) = \left\{ x : J \to \mathbb{R} : t \to (\psi(t) - \psi(a))^{1-\xi} x(t) \in C(\bar{J}, \mathbb{R}) \right\},$$

$\xi = \frac{1}{k}(\beta(k - \alpha) + \alpha)$

$$\mathcal{L} = \frac{\eta_1 (\psi(b) - \psi(a))^{\frac{\alpha}{k}}}{1 - \eta_2} \left[\frac{|c_2|}{|c_1 + c_2| \Gamma_k(k + \alpha)} + \frac{\Gamma_k(k\xi)}{\Gamma_k(\alpha + k\xi)} \right] < 1,$$

$$c_1 \left(\mathcal{J}_{a+}^{k(1-\xi), k; \psi} x \right)(a^+) + c_2 \left(\mathcal{J}_{a+}^{k(1-\xi), k; \psi} x \right)(b) = c_3.$$

Example 3.38 Taking $\beta \to 0, \alpha = \frac{1}{2}, k = 1, \psi(t) = t, c_1 = 1, c_2 = 0, c_3 = 0,$ and $\xi = \frac{1}{2},$ we obtain an initial valued problem which is a particular case of problem (3.49)–(3.50) with the Riemann-Liouville fractional derivative, given by

$$\left({}^{H}_{1}\mathcal{D}^{\frac{1}{2},0;\psi}_{1^{+}}x\right)(t) = \left({}^{RL}\mathcal{D}^{\frac{1}{2}}_{1^{+}}x\right)(t) = f\left(t, x(t), \left({}^{RL}\mathcal{D}^{\frac{1}{2}}_{1^{+}}x\right)(t)\right), \quad t \in (1, 2], \quad (3.64)$$

$$\left(\mathcal{J}^{\frac{1}{2},1;\psi}_{1^{+}}x\right)(1^{+}) = 0. \tag{3.65}$$

We have

$$C_{\xi;\psi}(J) = C_{\frac{1}{2};\psi}(J) = \left\{u : (1, 2] \to \mathbb{R} : (\sqrt{t-1})u \in C(\bar{J}, \mathbb{R})\right\},$$

and

$$C^{1}_{\xi;\psi}(J) = C^{1}_{\frac{1}{2};\psi}(J) = \left\{u \in C_{\frac{1}{2};\psi}(J) : u' \in C_{\frac{1}{2};\psi}(J)\right\}.$$

Since the continuous function $f \in C^{1}_{\frac{1}{2};\psi}(J)$, then the condition (3.32.1) is satisfied. For each $x, \bar{x}, y, \bar{y} \in \mathbb{R}$ and $t \in \bar{J}$, we have

$$|f(t, x, \bar{x}) - f(t, y, \bar{y})| \le \frac{1}{107e^{-t+2}} (|x - \bar{x}| + |y - \bar{y}|),$$

and so the condition (3.32.2) is satisfied with $\eta_1 = \eta_2 = \dfrac{1}{107}$. Also, the condition (3.61) of Theorem 3.32 is satisfied. Indeed, we have

$$\mathcal{L} = \frac{\sqrt{\pi}}{106} \approx 0.01672126 < 1.$$

Then the problem (3.64)–(3.65) has a unique solution in $C^{1}_{\frac{1}{2};\psi}([1, 2])$ and is Mittag-Leffler-Ulam-Hyers stable with respect to $\mathbb{E}^{\frac{1}{2}}_{1}\left(\sqrt{t-1}\right)$.

Example 3.39 Taking $\beta \to 1, \alpha = \frac{1}{2}, k = 1, \psi(t) = t, c_1 = 0, c_2 = 1, c_3 = 0$, and $\xi = 1$, we obtain a terminal value problem which is a particular case of problem (3.49)–(3.50) with Caputo fractional derivative, given by

$$\left({}^{H}_{1}\mathcal{D}^{\frac{1}{2},1;\psi}_{1^{+}}x\right)(t) = \left({}^{C}\mathcal{D}^{\frac{1}{2}}_{1^{+}}x\right)(t) = f\left(t, x(t), \left({}^{C}\mathcal{D}^{\frac{1}{2}}_{1^{+}}x\right)(t)\right), \quad t \in (1, 2], \quad (3.66)$$

$$\left(\mathcal{J}^{0,1;\psi}_{1^{+}}x\right)(2) = x(2) = 0. \tag{3.67}$$

We have $C_{\xi;\psi}(J) = C_{1;\psi}(J) = C(\bar{J}, \mathbb{R})$ and $C^{1}_{\xi;\psi}(J) = C^{1}_{1;\psi}(J) = C^{1}(\bar{J}, \mathbb{R})$. Also,

$$\mathcal{L} = \frac{4}{106\sqrt{\pi}} \approx 0.02129017 < 1.$$

As all the assumptions of Theorems 3.32 and 3.37 are satisfied, then the problem (3.66)–(3.67) has a unique solution in $C^1(\bar{J}, \mathbb{R})$ and is Mittag-Leffler-Ulam-Hyers stable with respect to $\mathbb{E}_1^{\frac{1}{2}}\left(\sqrt{t-1}\right)$.

Example 3.40 Taking $\beta \to \frac{1}{2}, \alpha = \frac{1}{2}, k = 1, \psi(t) = t, c_1 = 1, c_2 = 1, c_3 = 0$, and $\xi = \frac{3}{4}$, we obtain an anti-periodic problem which is a particular case of problem (3.49)–(3.50) with Hilfer fractional derivative, given by

$$\left({}^H\mathcal{D}_{1+}^{\frac{1}{2},\frac{1}{2};\psi}x\right)(t) = \left({}^H\mathcal{D}_{1+}^{\frac{1}{2},\frac{1}{2}}x\right)(t) = f\left(t, x(t), \left({}^H\mathcal{D}_{1+}^{\frac{1}{2},\frac{1}{2}}x\right)(t)\right), \quad t \in (1,2], \quad (3.68)$$

$$\left(\mathcal{J}_{1+}^{\frac{1}{4},1;\psi}x\right)(1) = -\left(\mathcal{J}_{1+}^{\frac{1}{4},1;\psi}x\right)(2). \quad (3.69)$$

We have

$$C_{\xi;\psi}(J) = C_{\frac{3}{4};\psi}(J) = \left\{u : (1,2] \to \mathbb{R} : (t-1)^{\frac{1}{4}}u \in C(\bar{J}, \mathbb{R})\right\},$$

and

$$C_{\xi;\psi}^1(J) = C_{\frac{1}{4};\psi}^1(J) = \left\{u \in C_{\frac{3}{4};\psi}(J) : u' \in C_{\frac{3}{4};\psi}(J)\right\}.$$

Also,

$$\mathcal{L} = \frac{1}{166}\left[\frac{1}{\sqrt{\pi}} + \frac{\Gamma(\frac{3}{4})}{\Gamma(\frac{5}{4})}\right] \approx 0.01154306579 < 1.$$

As all the assumptions of Theorem 3.32 and Theorem 3.37 are satisfied, then the problem (3.68)–(3.69) has a unique solution in $C_{\frac{1}{4};\psi}^1(J)$ and is Mittag-Leffler-Ulam-Hyers stable with respect to $\mathbb{E}_1^{\frac{1}{2}}\left(\sqrt{t-1}\right)$.

Example 3.41 Taking $\beta \to 0, \alpha = \frac{1}{2}, k = 1, \psi(t) = \ln(t), c_1 = 1, c_2 = 1, c_3 = 1$ and $\xi = \frac{1}{2}$, we obtain a boundary valued problem which is a particular case of problem (3.49)–(3.50) with Hadamard fractional derivative, given by

$$\left({}^H_1\mathcal{D}_{1+}^{\frac{1}{2},0;\psi}x\right)(t) = \left({}^H\mathcal{D}_{1+}^{\frac{1}{2}}x\right)(t) = f\left(t, x(t), \left({}^H\mathcal{D}_{1+}^{\frac{1}{2}}x\right)(t)\right), \quad t \in (1,2], \quad (3.70)$$

$$\left(\mathcal{J}_{1+}^{\frac{1}{2},1;\psi}x\right)(1) + \left(\mathcal{J}_{1+}^{\frac{1}{2},1;\psi}x\right)(2) = 1. \quad (3.71)$$

We have

$$C_{\xi;\psi}(J) = C_{\frac{1}{2};\psi}(J) = \left\{u : (1,2] \to \mathbb{R} : \sqrt{\ln(t)}u \in C(\bar{J}, \mathbb{R})\right\},$$

and

$$C_{\xi;\psi}^1(J) = C_{\frac{1}{2};\psi}^1(J) = \left\{u \in C_{\frac{1}{2};\psi}(J) : u' \in C_{\frac{1}{2};\psi}(J)\right\}.$$

Also

$$\mathcal{L} = \frac{\sqrt{\ln(2)}}{166} \left[\frac{1}{\sqrt{\pi}} + \sqrt{\pi} \right] \approx 0.011719176301 < 1.$$

As all the assumptions of Theorems 3.32 and 3.37 are satisfied, then the problem 3.70)–(3.71)
has a unique solution in $C^1_{\frac{1}{2};\psi}(J)$ and is Mittag-Leffler-Ulam-Hyers stable with respect to

$\mathbb{E}^{\frac{1}{2}}_1 \left(\sqrt{\ln(t)} \right).$

Remark 3.42 By varying β and the function ψ, we can obtain several cases of our problem
(3.49)–(3.50). And if we take the same steps as the last examples with appropriate conditions,
we can prove the existence, uniqueness, and Mittag-Leffler-Ulam-Hyers stability results for
each case.

We may have additional problems with the following fractional derivative :

- Caputo-Hadmard derivative: By taking $\beta \to 1$, $k = 1$, $\psi(t) = \ln(t)$.
- Hilfer-Hadmard derivative: By taking $\beta \in (0, 1)$, $k = 1$, $\psi(t) = \ln(t)$.
- Katugampola derivative: By taking $\beta \to 0$, $k = 1$, $\psi(t) = t^\rho$.
- Caputo-Katugampola derivative: By taking $\beta \to 1$, $k = 1$, $\psi(t) = t^\rho$.
- Hilfer-Katugampola derivative: By taking $\beta \in (0, 1)$, $k = 1$, $\psi(t) = t^\rho$.

3.6   Notes and Remarks

The results of this chapter are taken from Salim et al. [126, 127, 134, 137]. For more relevant
results and studies, one can see the monographs [7, 14, 23, 27, 37, 50, 62, 68, 85, 100] and
the papers [57, 63, 65, 90, 91, 102–104, 122, 124, 129, 131, 143–149].

Fractional Differential Equations with Instantaneous Impulses

4

4.1   Introduction and Motivations

The aim of this chapter is to prove some existence, uniqueness, and Ulam-Hyers-Rassias stability results for a class of boundary value problem for nonlinear implicit fractional differential equations with impulses and generalized Hilfer-type fractional derivative. We base our arguments on some relevant fixed point theorems combined with the technique of measure of noncompactness. Examples are included to show the applicability of our results for each section.

The outcome of our study in this chapter can be considered as a partial continuation of the problems raised recently in the following:

- The monographs of Abbas et al. [7, 8, 14], Baleanu et al. [43], and Rassias et al. [115], and the papers of Afshari et al. [20, 21], Benchohra et al. [48, 49], Karapınar et al. [17, 19, 34, 83, 84], and Zhou et al. [162], which deal with various linear and nonlinear initial and boundary value problems for fractional differential equations involving different kinds of fractional derivatives.
- The monographs of Benchohra et al. [50], Graef et al. [73], and Samoilenko et al. [139], and the papers of Abbas et al. [9] and Benchohra et al. [49] where the authors investigated various problems with fractional differential equations and impulsive conditions.
- The monographs of Abbas et al. [7, 13], and the papers of Abbas et al. [10, 12], Benchohra et al. [51–53], and Kucche et al. [89, 96, 141]; in it, considerable attention has been given to the study of the Ulam-Hyers and Ulam-Hyers-Rassias stability of various classes of functional equations.
- The paper of Harikrishnan et al. [76]; in it, the authors investigated existence theory and different kinds of stability in the sense of Ulam, for the following boundary value problem with nonlinear generalized Hilfer-type fractional differential equation with impulses:

© The Author(s), under exclusive license to Springer Nature Switzerland AG 2023   77
M. Benchohra et al., *Advanced Topics in Fractional Differential Equations*,
Synthesis Lectures on Mathematics & Statistics,
https://doi.org/10.1007/978-3-031-26928-8_4

$$\begin{cases} \left({}^\rho \mathcal{D}^{\alpha,\beta} u\right)(t) = f(t, u(t)); \ t \in \bar{I} := I \setminus \{t_1, \ldots, t_m\}, \ I := [0, b], \\ \Delta^\rho \mathcal{J}^{1-\gamma} u(t) \big|_{t=t_k} = L_k(u(t_k^-)); k = 1, \ldots, m, \\ {}^\rho \mathcal{J}^{1-\gamma} u(0) = u_0, \end{cases}$$

where ${}^\rho \mathcal{D}^{\alpha,\beta}, {}^\rho \mathcal{J}^{1-\gamma}$ are the generalized Hilfer fractional derivative of order $\alpha \in (0, 1)$ and type $\beta \in [0, 1]$ and generalized fractional integral of order $1 - \gamma$, $(\gamma = \alpha + \beta - \alpha\beta)$, respectively, $0 = t_0 < t_1 < \ldots < t_m < t_{m+1} = b < \infty$, $u(t_k^+) = \lim\limits_{\epsilon \to 0^+} u(t_k + \epsilon)$ and $u(t_k^-) = \lim\limits_{\epsilon \to 0^-} u(t_k + \epsilon)$ represent the right- and left-hand limits of $u(t)$ at $t = t_k$, $\Delta^\rho \mathcal{J}^{1-\gamma} u(t) \big|_{t=t_k} = {}^\rho \mathcal{J}^{1-\gamma} u(t_k^+) - {}^\rho \mathcal{J}^{1-\gamma} u(t_k^-)$, $f : I \times \mathbb{R} \to \mathbb{R}$ is a given function, and $L_k : \mathbb{R} \to \mathbb{R}; k = 1, \ldots, m$ are given continuous functions.

4.2   Existence and Ulam Stability Results for Generalized Hilfer-Type Boundary Value Problem

In this section, we establish the existence and uniqueness results to the boundary value problem with nonlinear implicit generalized Hilfer-type fractional differential equation with impulses:

$$\left({}^\rho \mathcal{D}^{\alpha,\beta}_{t_k^+} u\right)(t) = f\left(t, u(t), \left({}^\rho \mathcal{D}^{\alpha,\beta}_{t_k^+} u\right)(t)\right); \ t \in J_k, \ k = 0, \ldots, m, \qquad (4.1)$$

$$\left({}^\rho \mathcal{J}^{1-\gamma}_{t_k^+} u\right)(t_k^+) = \left({}^\rho \mathcal{J}^{1-\gamma}_{t_{k-1}^+} u\right)(t_k^-) + L_k(u(t_k^-)); k = 1, \ldots, m, \qquad (4.2)$$

$$c_1 \left({}^\rho \mathcal{J}^{1-\gamma}_{a^+} u\right)(a^+) + c_2 \left({}^\rho \mathcal{J}^{1-\gamma}_{t_m^+} u\right)(b) = c_3, \qquad (4.3)$$

where ${}^\rho \mathcal{D}^{\alpha,\beta}_{t_k^+}, {}^\rho \mathcal{J}^{1-\gamma}_{t_k^+}$ are the generalized Hilfer fractional derivative of order $\alpha \in (0, 1)$ and type $\beta \in [0, 1]$ and generalized fractional integral of order $1 - \gamma$, $(\gamma = \alpha + \beta - \alpha\beta), \rho > 0$ respectively, c_1, c_2, c_3 are reals with $c_1 + c_2 \neq 0$, $J_k := (t_k, t_{k+1}]; k = 0, \ldots, m, a = t_0 < t_1 < \ldots < t_m < t_{m+1} = b < \infty, u(t_k^+) = \lim\limits_{\epsilon \to 0^+} u(t_k + \epsilon)$ and $u(t_k^-) = \lim\limits_{\epsilon \to 0^-} u(t_k + \epsilon)$ represent the right- and left-hand limits of $u(t)$ at $t = t_k$, $f : J \times \mathbb{R} \times \mathbb{R} \to \mathbb{R}$ is a given function, and $L_k : \mathbb{R} \to \mathbb{R}; k = 1, \ldots, m$ are given continuous functions.

4.2.1   Existence Results

Before establishing our existence results, we need to define the following weighted Banach space:

$$PC_{\gamma,\rho}(J) = \Big\{ u : J \to \mathbb{R} : u(t) \in C_{\gamma,\rho}(J_k); k = 0, \ldots, m, \text{ and there exist}$$

$$u(t_k^-) \text{ and } \left({}^{\rho}\mathcal{J}_{t_k^+}^{1-\gamma} u \right)(t_k^+); k = 0, \ldots, m, \text{ with } u(t_k^-) = u(t_k) \Big\},$$

and

$$PC_{\gamma,\rho}^n(J) = \Big\{ u \in PC^{n-1} : u^{(n)} \in PC_{\gamma,\rho}(J) \Big\}, n \in \mathbb{N},$$

$$PC_{\gamma,\rho}^0(J) = PC_{\gamma,\rho}(J),$$

with the norm

$$\|u\|_{PC_{\gamma,\rho}} = \max_{k=0,\ldots,m} \left\{ \sup_{t \in [t_k, t_{k+1}]} \left| \left(\frac{t^{\rho} - t_k^{\rho}}{\rho} \right)^{1-\gamma} u(t) \right| \right\}.$$

We also define the space

$$PC_{\gamma,\rho}^{\gamma}(J) = \Big\{ u \in PC_{\gamma,\rho}(J), {}^{\rho}\mathcal{D}_{t_k^+}^{\gamma} u \in PC_{\gamma,\rho}(J) \Big\}, k = 0, \ldots, m.$$

Let us now consider the following linear fractional differential equation:

$$\left({}^{\rho}\mathcal{D}_{t_k^+}^{\alpha,\beta} u \right)(t) = \psi(t), \quad t \in J_k, k = 0, \ldots, m, \tag{4.4}$$

where $0 < \alpha < 1, 0 \le \beta \le 1, \rho > 0$, with the conditions

$$\left({}^{\rho}\mathcal{J}_{t_k^+}^{1-\gamma} u \right)(t_k^+) = \left({}^{\rho}\mathcal{J}_{t_{k-1}^+}^{1-\gamma} u \right)(t_k^-) + L_k(u(t_k^-)); k = 1, \ldots, m, \tag{4.5}$$

and

$$c_1 \left({}^{\rho}\mathcal{J}_{a^+}^{1-\gamma} u \right)(a^+) + c_2 \left({}^{\rho}\mathcal{J}_{t_m^+}^{1-\gamma} u \right)(b) = c_3, \tag{4.6}$$

where $\gamma = \alpha + \beta - \alpha\beta, c_1, c_2, c_3 \in \mathbb{R}$ with

$$c_1 + c_2 \ne 0, \quad \vartheta_1 = \frac{c_2}{c_1 + c_2}, \quad \vartheta_2 = \frac{c_3}{c_1 + c_2}$$

and

$$p^* = \sup\left\{\left(\frac{t_k^\rho - t_{k-1}^\rho}{\rho}\right)^{\gamma-1} : k = 1, \ldots, m\right\},$$

such that $\psi : J \to \mathbb{R}$ be a function satisfying the functional equation

$$\psi(t) = f(t, u(t), \psi(t)).$$

The following theorem shows that the problem (4.4)–(4.6) has a unique solution given by

$$u(t) = \begin{cases} \dfrac{\left(\dfrac{t^\rho - a^\rho}{\rho}\right)^{\gamma-1}}{\Gamma(\gamma)}\left[\vartheta_2 - \vartheta_1 \displaystyle\sum_{i=1}^{m} L_i(u(t_i^-)) - \vartheta_1 \sum_{i=1}^{m}\left({}^\rho\mathcal{J}_{(t_{i-1})^+}^{1-\gamma+\alpha}\psi\right)(t_i)\right. \\ \quad \left. - \vartheta_1\left({}^\rho\mathcal{J}_{t_m^+}^{1-\gamma+\alpha}\psi\right)(b)\right] + \displaystyle\int_a^t \left(\dfrac{t^\rho - s^\rho}{\rho}\right)^{\alpha-1}\dfrac{s^{\rho-1}\psi(s)ds}{\Gamma(\alpha)}, \quad t \in J_0, \\[4ex] \dfrac{\left(\dfrac{t^\rho - t_k^\rho}{\rho}\right)^{\gamma-1}}{\Gamma(\gamma)}\left[\vartheta_2 - \vartheta_1 \displaystyle\sum_{i=1}^{m} L_i(u(t_i^-)) - \vartheta_1 \sum_{i=1}^{m}\left({}^\rho\mathcal{J}_{(t_{i-1})^+}^{1-\gamma+\alpha}\psi\right)(t_i)\right. \\ \quad \left. - \vartheta_1\left({}^\rho\mathcal{J}_{t_m^+}^{1-\gamma+\alpha}\psi\right)(b) + \displaystyle\sum_{i=1}^{k} L_i(u(t_i^-)) + \sum_{i=1}^{k}\left({}^\rho\mathcal{J}_{(t_{i-1})^+}^{1-\gamma+\alpha}\psi\right)(t_i)\right] \\ \quad + \left({}^\rho\mathcal{J}_{t_k^+}^{\alpha}\psi\right)(t), \quad t \in J_k, k = 1, \ldots, m. \end{cases} \tag{4.7}$$

Theorem 4.1 *Let* $\gamma = \alpha + \beta - \alpha\beta$, *where* $0 < \alpha < 1$ *and* $0 \le \beta \le 1$. *If* $\psi : J \to \mathbb{R}$ *is a function such that* $\psi(\cdot) \in PC_{\gamma,\rho}(J)$, *then* $u \in PC_{\gamma,\rho}^{\gamma}(J)$ *satisfies the problem (4.4)–(4.6) if and only if it satisfies (4.7).*

Proof Assume u satisfies (4.4)–(4.6). If $t \in J_0$, then

$$\left({}^\rho\mathcal{D}_{a^+}^{\alpha,\beta}u\right)(t) = \psi(t).$$

Lemma 2.38 implies we have a solution that can be written as

$$u(t) = \frac{\left({}^\rho\mathcal{J}_{a^+}^{1-\gamma}u\right)(a^+)}{\Gamma(\gamma)}\left(\frac{t^\rho - a^\rho}{\rho}\right)^{\gamma-1} + \frac{1}{\Gamma(\alpha)}\int_a^t \left(\frac{t^\rho - s^\rho}{\rho}\right)^{\alpha-1}s^{\rho-1}\psi(s)ds. \tag{4.8}$$

If $t \in J_1$, then Lemma 2.38 implies

$$u(t) = \frac{\left(^{\rho}\mathcal{J}_{t_1^+}^{1-\gamma} u\right)(t_1^+)}{\Gamma(\gamma)} \left(\frac{t^{\rho} - t_1^{\rho}}{\rho}\right)^{\gamma-1} + \frac{1}{\Gamma(\alpha)} \int_{t_1}^{t} \left(\frac{t^{\rho} - s^{\rho}}{\rho}\right)^{\alpha-1} s^{\rho-1} \psi(s) ds$$

$$= \frac{\left(^{\rho}\mathcal{J}_{a^+}^{1-\gamma} u\right)(t_1^-) + L_1(u(t_1^-))}{\Gamma(\gamma)} \left(\frac{t^{\rho} - t_1^{\rho}}{\rho}\right)^{\gamma-1} + \left(^{\rho}\mathcal{J}_{t_1^+}^{\alpha} \psi\right)(t)$$

$$= \frac{(t^{\rho} - t_1^{\rho})^{\gamma-1}}{\Gamma(\gamma)\rho^{\gamma-1}} \left[\left(^{\rho}\mathcal{J}_{a^+}^{1-\gamma} u\right)(a^+) + L_1(u(t_1^-)) + \left(^{\rho}\mathcal{J}_{a^+}^{1-\gamma+\alpha} \psi\right)(t_1) \right]$$

$$+ \left(^{\rho}\mathcal{J}_{t_1^+}^{\alpha} \psi\right)(t).$$

If $t \in J_2$, then Lemma 2.38 implies

$$u(t) = \frac{\left(^{\rho}\mathcal{J}_{t_2^+}^{1-\gamma} u\right)(t_2^+)}{\Gamma(\gamma)} \left(\frac{t^{\rho} - t_2^{\rho}}{\rho}\right)^{\gamma-1} + \frac{1}{\Gamma(\alpha)} \int_{t_2}^{t} \left(\frac{t^{\rho} - s^{\rho}}{\rho}\right)^{\alpha-1} s^{\rho-1} \psi(s) ds$$

$$= \frac{\left(^{\rho}\mathcal{J}_{t_1^+}^{1-\gamma} u\right)(t_2^-) + L_2(u(t_2^-))}{\Gamma(\gamma)} \left(\frac{t^{\rho} - t_2^{\rho}}{\rho}\right)^{\gamma-1} + \left(^{\rho}\mathcal{J}_{t_2^+}^{\alpha} \psi\right)(t)$$

$$= \frac{1}{\Gamma(\gamma)} \left(\frac{t^{\rho} - t_2^{\rho}}{\rho}\right)^{\gamma-1} \left[\left(^{\rho}\mathcal{J}_{a^+}^{1-\gamma} u\right)(a^+) + L_1(u(t_1^-)) + L_2(u(t_2^-)) \right.$$

$$\left. + \left(^{\rho}\mathcal{J}_{a^+}^{1-\gamma+\alpha} \psi\right)(t_1) + \left(^{\rho}\mathcal{J}_{t_1^+}^{1-\gamma+\alpha} \psi\right)(t_2) \right] + \left(^{\rho}\mathcal{J}_{t_2^+}^{\alpha} \psi\right)(t).$$

Repeating the process in this way, the solution $u(t)$ for $t \in J_k$, $k = 1, \ldots, m$, can be written as

$$u(t) = \left[\left(^{\rho}\mathcal{J}_{a^+}^{1-\gamma} u\right)(a^+) + \sum_{i=1}^{k} L_i(u(t_i^-)) + \sum_{i=1}^{k} \left(^{\rho}\mathcal{J}_{(t_{i-1})^+}^{1-\gamma+\alpha} \psi\right)(t_i) \right]$$

$$\times \frac{1}{\Gamma(\gamma)} \left(\frac{t^{\rho} - t_k^{\rho}}{\rho}\right)^{\gamma-1} + \left(^{\rho}\mathcal{J}_{t_k^+}^{\alpha} \psi\right)(t). \tag{4.9}$$

Applying $^{\rho}\mathcal{J}_{t_m^+}^{1-\gamma}$ on both sides of (4.9), using Lemma 2.19 and taking $t = b$, we obtain

$$\left(^{\rho}\mathcal{J}_{t_m^+}^{1-\gamma} u\right)(b) = \left(^{\rho}\mathcal{J}_{a^+}^{1-\gamma} u\right)(a^+) + \sum_{i=1}^{m} L_i(u(t_i^-)) + \sum_{i=1}^{m} \left(^{\rho}\mathcal{J}_{(t_{i-1})^+}^{1-\gamma+\alpha} \psi\right)(t_i)$$

$$+ \left(^{\rho}\mathcal{J}_{(t_m)^+}^{1-\gamma+\alpha} \psi\right)(b). \tag{4.10}$$

Multiplying both sides of (4.10) by c_2 and using condition (4.6), we obtain

$$c_3 - c_1 \left({}^{\rho}\mathcal{J}_{a^+}^{1-\gamma} u \right)(a^+) = c_2 \left({}^{\rho}\mathcal{J}_{a^+}^{1-\gamma} u \right)(a^+) + c_2 \sum_{i=1}^{m} L_i(u(t_i^-))$$

$$+ c_2 \sum_{i=1}^{m} \left({}^{\rho}\mathcal{J}_{(t_{i-1})^+}^{1-\gamma+\alpha} \psi \right)(t_i) + c_2 \left({}^{\rho}\mathcal{J}_{(t_m)^+}^{1-\gamma+\alpha} \psi \right)(b),$$

which implies that

$$\left({}^{\rho}\mathcal{J}_{a^+}^{1-\gamma} u \right)(a^+) = \vartheta_2 - \vartheta_1 \sum_{i=1}^{m} L_i(u(t_i^-)) - \vartheta_1 \sum_{i=1}^{m} \left({}^{\rho}\mathcal{J}_{(t_{i-1})^+}^{1-\gamma+\alpha} \psi \right)(t_i)$$

$$- \vartheta_1 \left({}^{\rho}\mathcal{J}_{(t_m)^+}^{1-\gamma+\alpha} \psi \right)(b). \tag{4.11}$$

Substituting (4.11) into (4.9) and (4.8), we obtain (4.7).
Reciprocally, applying ${}^{\rho}\mathcal{J}_{t_k^+}^{1-\gamma}$ on both sides of (4.7) and using Lemma 2.19 and Theorem 2.14, we get

$$\left({}^{\rho}\mathcal{J}_{t_k^+}^{1-\gamma} u \right)(t) = \begin{cases} \vartheta_2 - \vartheta_1 \sum_{i=1}^{m} L_i(u(t_i^-)) - \vartheta_1 \sum_{i=1}^{m} \left({}^{\rho}\mathcal{J}_{(t_{i-1})^+}^{1-\gamma+\alpha} \psi \right)(t_i) \\ \quad - \vartheta_1 \left({}^{\rho}\mathcal{J}_{(t_m)^+}^{1-\gamma+\alpha} \psi \right)(b) + \left({}^{\rho}\mathcal{J}_{a^+}^{1-\gamma+\alpha} \psi \right)(t), \ t \in J_0, \\[4mm] \vartheta_2 - \vartheta_1 \sum_{i=1}^{m} L_i(u(t_i^-)) - \vartheta_1 \sum_{i=1}^{m} \left({}^{\rho}\mathcal{J}_{(t_{i-1})^+}^{1-\gamma+\alpha} \psi \right)(t_i) \\ \quad - \vartheta_1 \left({}^{\rho}\mathcal{J}_{t_m^+}^{1-\gamma+\alpha} \psi \right)(b) + \sum_{i=1}^{k} \left({}^{\rho}\mathcal{J}_{(t_{i-1})^+}^{1-\gamma+\alpha} \psi \right)(t_i) \\ \quad + \sum_{i=1}^{k} L_i(u(t_i^-)) + \left({}^{\rho}\mathcal{J}_{t_k^+}^{1-\gamma+\alpha} \psi \right)(t), \ t \in J_k, k \neq 0. \end{cases} \tag{4.12}$$

Next, taking the limit $t \to a^+$ of (4.12) and using Lemma 2.24, with $1 - \gamma < 1 - \gamma + \alpha$, we obtain

$$\left({}^{\rho}\mathcal{J}_{a^+}^{1-\gamma} u \right)(a^+) = \vartheta_2 - \vartheta_1 \sum_{i=1}^{m} L_i(u(t_i^-)) - \vartheta_1 \sum_{i=1}^{m} \left({}^{\rho}\mathcal{J}_{(t_{i-1})^+}^{1-\gamma+\alpha} \psi \right)(t_i)$$

$$- \vartheta_1 \left({}^{\rho}\mathcal{J}_{t_m^+}^{1-\gamma+\alpha} \psi \right)(b). \tag{4.13}$$

Now, taking $t = b$ in (4.12), we get

$$\left({}^{\rho}\mathcal{J}_{t_m^+}^{1-\gamma} u \right)(b) = \vartheta_2 + (1 - \vartheta_1) \left(\sum_{i=1}^{m} L_i(u(t_i^-)) + \sum_{i=1}^{m} \left({}^{\rho}\mathcal{J}_{(t_{i-1})^+}^{1-\gamma+\alpha} \psi \right)(t_i) \right.$$

$$\left. + \left({}^{\rho}\mathcal{J}_{t_m^+}^{1-\gamma+\alpha} \psi \right)(b) \right). \tag{4.14}$$

From (4.13) and (4.14), we find that

$$c_1 \left({}^{\rho}\mathcal{J}_{a^+}^{1-\gamma} u \right)(a^+) + c_2 \left({}^{\rho}\mathcal{J}_{t_m^+}^{1-\gamma} u \right)(b) = c_3,$$

which shows that the boundary condition $c_1 \left({}^{\rho}\mathcal{J}_{a^+}^{1-\gamma} u \right)(a^+) + c_2 \left({}^{\rho}\mathcal{J}_{t_m^+}^{1-\gamma} u \right)(b) = c_3$ is satisfied.

Next, apply operator ${}^{\rho}\mathcal{D}_{t_k^+}^{\gamma}$ on both sides of (4.7), where $k = 0, \ldots, m$. Then, from Lemma 2.19 and Lemma 2.33, we obtain

$$({}^{\rho}\mathcal{D}_{t_k^+}^{\gamma} u)(t) = \left({}^{\rho}\mathcal{D}_{t_k^+}^{\beta(1-\alpha)} \psi \right)(t). \tag{4.15}$$

Since $u \in C_{\gamma,\rho}^{\gamma}(J_k)$ and by definition of $C_{\gamma,\rho}^{\gamma}(J_k)$, we have ${}^{\rho}\mathcal{D}_{t_k^+}^{\gamma} u \in C_{\gamma,\rho}(J_k)$, then (4.15) implies that

$$({}^{\rho}\mathcal{D}_{t_k^+}^{\gamma} u)(t) = \left(\delta_{\rho} \; {}^{\rho}\mathcal{J}_{t_k^+}^{1-\beta(1-\alpha)} \psi \right)(t) = \left({}^{\rho}\mathcal{D}_{t_k^+}^{\beta(1-\alpha)} \psi \right)(t) \in C_{\gamma,\rho}(J_k). \tag{4.16}$$

As $\psi(\cdot) \in C_{\gamma,\rho}(J_k)$ and from Lemma 2.23, follows

$$\left({}^{\rho}\mathcal{J}_{t_k^+}^{1-\beta(1-\alpha)} \psi \right) \in C_{\gamma,\rho}(J_k). \tag{4.17}$$

From (4.16), (4.17), and by the definition of the space $C_{\gamma,\rho}^{n}(J_k)$, we obtain

$$\left({}^{\rho}\mathcal{J}_{t_k^+}^{1-\beta(1-\alpha)} \psi \right) \in C_{\gamma,\rho}^{1}(J_k).$$

Applying operator ${}^{\rho}\mathcal{J}_{t_k^+}^{\beta(1-\alpha)}$ on both sides of (4.15) and using Lemma 2.32, Lemma 2.24, and Property 2.22, we have

$$\left({}^{\rho}\mathcal{D}_{t_k^+}^{\alpha,\beta} u \right)(t) = {}^{\rho}\mathcal{J}_{t_k^+}^{\beta(1-\alpha)} \left({}^{\rho}\mathcal{D}_{t_k^+}^{\gamma} u \right)(t)$$

$$= \psi(t) - \frac{\left({}^{\rho}\mathcal{J}_{t_k^+}^{1-\beta(1-\alpha)} \psi \right)(t_k)}{\Gamma(\beta(1-\alpha))} \left(\frac{t^{\rho} - t_k^{\rho}}{\rho} \right)^{\beta(1-\alpha)-1}$$

$$= \psi(t),$$

that is, (4.4) holds.

Also, we can easily show that

$$\left({}^{\rho}\mathcal{J}_{t_k^+}^{1-\gamma} u \right)(t_k^+) = \left({}^{\rho}\mathcal{J}_{t_{k-1}^+}^{1-\gamma} u \right)(t_k^-) + L_k(u(t_k^-)); \; k = 1, \ldots, m.$$

This completes the proof.   □

As a consequence of Theorem 4.1, we have the following result.

Lemma 4.2 *Let $\gamma = \alpha + \beta - \alpha\beta$ where $0 < \alpha < 1$ and $0 \leq \beta \leq 1$, let $f : J \times \mathbb{R} \times \mathbb{R} \to \mathbb{R}$ be a function such that $f(\cdot, u(\cdot), w(\cdot)) \in PC_{\gamma,\rho}(J)$ for any $u, w \in PC_{\gamma,\rho}(J)$. If $u \in PC_{\gamma,\rho}^{\gamma}(J)$, then u satisfies the problem (4.1)–(4.3) if and only if u is the fixed point of the operator $\Psi : PC_{\gamma,\rho}(J) \to PC_{\gamma,\rho}(J)$ defined by*

$$
\Psi u(t) = \frac{1}{\Gamma(\gamma)} \left(\frac{t^{\rho} - t_{k}^{\rho}}{\rho} \right)^{\gamma-1} \left[\vartheta_2 - \vartheta_1 \sum_{i=1}^{m} L_i(u(t_i^-)) - \vartheta_1 \sum_{i=1}^{m} \left({}^{\rho}\mathcal{J}_{(t_{i-1})^+}^{1-\gamma+\alpha} h \right)(t_i) \right.
$$

$$
\left. - \vartheta_1 \left({}^{\rho}\mathcal{J}_{t_m^+}^{1-\gamma+\alpha} h \right)(b) + \sum_{a<t_k<t} L_k(u(t_k^-)) + \sum_{a<t_k<t} \left({}^{\rho}\mathcal{J}_{(t_{k-1})^+}^{1-\gamma+\alpha} h \right)(t_k) \right]
$$

$$
+ \left({}^{\rho}\mathcal{J}_{t_k^+}^{\alpha} h \right)(t), \quad t \in J_k, k = 0, \dots, m, \tag{4.18}
$$

where $h : J \to \mathbb{R}$ be a function satisfying the functional equation

$$
h(t) = f(t, u(t), h(t)).
$$

We are now in a position to state and prove our existence result for problem (4.1)–(4.3) based on Banach's fixed point.

Theorem 4.3 *Assume that the following hypotheses hold.*

(4.3.1) The continuous function $f : J \times \mathbb{R} \times \mathbb{R} \to \mathbb{R}$ be such that

$$
f(\cdot, u(\cdot), w(\cdot)) \in PC_{\gamma,\rho}^{\beta(1-\alpha)}(J) \text{ for any } u, w \in PC_{\gamma,\rho}(J).
$$

(4.3.2) There exist constants $K > 0$ and $0 < M < 1$ such that

$$
|f(t, u, w) - f(t, \bar{u}, \bar{w})| \leq K|u - \bar{u}| + M|w - \bar{w}|
$$

for any $u, w, \bar{u}, \bar{w} \in \mathbb{R}$ and $t \in J$.

(4.3.3) There exists a constant $l^* > 0$ such that

$$
|L_k(u) - L_k(\bar{u})| \leq l^*|u - \bar{u}|
$$

for any $u, \bar{u} \in \mathbb{R}$ and $k = 1, \dots, m$.

If

$$L := \left(|\vartheta_1| + 1\right)\left(\frac{ml^* p^*}{\Gamma(\gamma)} + \frac{mK}{(1 - M)\Gamma(1 + \alpha)}\left(\frac{b^\rho - a^\rho}{\rho}\right)^\alpha\right)$$
$$+ \frac{K}{(1 - M)}\left(\frac{|\vartheta_1|}{\Gamma(1 + \alpha)} + \frac{\Gamma(\gamma)}{\Gamma(\gamma + \alpha)}\right)\left(\frac{b^\rho - a^\rho}{\rho}\right)^\alpha < 1, \tag{4.19}$$

then the problem (4.1)–(4.3) has a unique solution in $PC_{\gamma,\rho}^\gamma(J)$.

Proof The proof will be given in two steps.

Step 1: We show that the operator Ψ defined in (4.18) has a unique fixed point u^* in $PC_{\gamma,\rho}(J)$. Let $u, w \in PC_{\gamma,\rho}(J)$ and $t \in J$, then we have

$$|\Psi u(t) - \Psi w(t)|$$

$$\leq \left[|\vartheta_1|\sum_{i=1}^m |L_i(u(t_i^-)) - L_i(w(t_i^-))| + |\vartheta_1|\left({}^\rho\mathcal{J}_{t_m^+}^{1-\gamma+\alpha}|h(s) - g(s)|\right)(b)\right.$$

$$+ |\vartheta_1|\sum_{i=1}^m \left({}^\rho\mathcal{J}_{(t_{i-1})^+}^{1-\gamma+\alpha}|h(s) - g(s)|\right)(t_i) + \sum_{a < t_k < t} |L_k(u(t_k^-)) - L_k(w(t_k^-))|$$

$$\left. + \sum_{a < t_k < t}\left({}^\rho\mathcal{J}_{(t_{k-1})^+}^{1-\gamma+\alpha}|h(s) - g(s)|\right)(t_k)\right]\frac{(t^\rho - t_k^\rho)^{\gamma-1}}{\Gamma(\gamma)\rho^{\gamma-1}}$$

$$+ \left({}^\rho\mathcal{J}_{t_k^+}^\alpha|h(s) - g(s)|\right)(t),$$

where $h, g \in PC_{\gamma,\rho}(J)$ such that

$$h(t) = f(t, u(t), h(t)),$$
$$g(t) = f(t, w(t), g(t)).$$

By (4.3.2), we have

$$|h(t) - g(t)| = |f(t, u(t), h(t)) - f(t, w(t), g(t))|$$
$$\leq K|u(t) - w(t)| + M|h(t) - g(t)|.$$

Then,

$$|h(t) - g(t)| \leq \frac{K}{1 - M}|u(t) - w(t)|.$$

Therefore, for each $t \in J$

$$|\Psi u(t) - \Psi w(t)|$$

$$\leq \left[|\vartheta_1| \sum_{i=1}^{m} l^* |u(t_i) - w(t_i)| + \frac{|\vartheta_1| K}{1-M} \left({}^{\rho}\mathcal{J}_{t_m^+}^{1-\gamma+\alpha} |u(s) - w(s)| \right) (b) \right.$$

$$+ \frac{|\vartheta_1| K}{1-M} \sum_{i=1}^{m} \left({}^{\rho}\mathcal{J}_{(t_{i-1})^+}^{1-\gamma+\alpha} |u(s) - w(s)| \right) (t_i) + \sum_{i=1}^{m} l^* |u(t_i) - w(t_i)|$$

$$+ \frac{K}{1-M} \sum_{i=1}^{m} \left({}^{\rho}\mathcal{J}_{(t_{i-1})^+}^{1-\gamma+\alpha} |u(s) - w(s)| \right) (t_i) \right] \frac{(t^\rho - t_k^\rho)^{\gamma-1}}{\Gamma(\gamma)\rho^{\gamma-1}}$$

$$+ \frac{K}{1-M} \left({}^{\rho}\mathcal{J}_{t_k^+}^{\alpha} |u(s) - w(s)| \right) (t).$$

Thus

$$|\Psi u(t) - \Psi w(t)|$$

$$\leq \frac{(t^\rho - t_k^\rho)^{\gamma-1}}{\Gamma(\gamma)\rho^{\gamma-1}} \left[|\vartheta_1| ml^* p^* + \frac{|\vartheta_1| K}{1-M} \left({}^{\rho}\mathcal{J}_{t_m^+}^{1-\gamma+\alpha} \left(\frac{s^\rho - t_m^\rho}{\rho} \right)^{\gamma-1} \right) (b) \right.$$

$$+ \frac{m K |\vartheta_1|}{1-M} \left({}^{\rho}\mathcal{J}_{(t_{k-1})^+}^{1-\gamma+\alpha} \left(\frac{s^\rho - t_{k-1}^\rho}{\rho} \right)^{\gamma-1} \right) (t_k)$$

$$+ ml^* p^* + \frac{m K}{1-M} \left({}^{\rho}\mathcal{J}_{(t_{k-1})^+}^{1-\gamma+\alpha} \left(\frac{s^\rho - t_{k-1}^\rho}{\rho} \right)^{\gamma-1} \right) (t_k) \right] \|u - w\|_{PC_{\gamma,\rho}}$$

$$+ \frac{K}{1-M} \|u - w\|_{PC_{\gamma,\rho}} \left({}^{\rho}\mathcal{J}_{t_k^+}^{\alpha} \left(\frac{s^\rho - t_k^\rho}{\rho} \right)^{\gamma-1} \right) (t).$$

By Lemma 2.19, we have

$$|\Psi u(t) - \Psi w(t)|$$

$$\leq \frac{1}{\Gamma(\gamma)} \left(\frac{t^\rho - t_k^\rho}{\rho} \right)^{\gamma-1} \|u - w\|_{PC_{\gamma,\rho}} \left[|\vartheta_1| ml^* p^* \right.$$

$$+ \frac{|\vartheta_1| K \Gamma(\gamma)}{(1-M)\Gamma(1+\alpha)} \left(\frac{b^\rho - t_m^\rho}{\rho} \right)^{\alpha} + \frac{m K |\vartheta_1| \Gamma(\gamma)}{(1-M)\Gamma(1+\alpha)} \left(\frac{t_k^\rho - t_{k-1}^\rho}{\rho} \right)^{\alpha}$$

$$+ ml^* p^* + \frac{m K \Gamma(\gamma)}{(1-M)\Gamma(1+\alpha)} \left(\frac{t_k^\rho - t_{k-1}^\rho}{\rho} \right)^{\alpha} \right]$$

$$+ \frac{K \Gamma(\gamma)}{(1-M)\Gamma(\gamma+\alpha)} \|u - w\|_{PC_{\gamma,\rho}} \left(\frac{t^\rho - t_k^\rho}{\rho} \right)^{\alpha+\gamma-1},$$

hence

$$\left| \left(\frac{t^\rho - t_k^\rho}{\rho} \right)^{1-\gamma} (\Psi u(t) - \Psi w(t)) \right|$$

$$\leq \left[(|\vartheta_1| + 1) \left(\frac{ml^* p^*}{\Gamma(\gamma)} + \frac{mK}{(1-M)\Gamma(1+\alpha)} \left(\frac{b^\rho - a^\rho}{\rho} \right)^\alpha \right) \right.$$
$$\left. + \frac{K}{(1-M)} \left(\frac{|\vartheta_1|}{\Gamma(1+\alpha)} + \frac{\Gamma(\gamma)}{\Gamma(\gamma+\alpha)} \right) \left(\frac{b^\rho - a^\rho}{\rho} \right)^\alpha \right] \|u - w\|_{PC_{\gamma,\rho}},$$

which implies that

$$\|\Psi u - \Psi w\|_{PC_{\gamma,\rho}}$$
$$\leq \left[(|\vartheta_1| + 1) \left(\frac{ml^* p^*}{\Gamma(\gamma)} + \frac{mK}{(1-M)\Gamma(1+\alpha)} \left(\frac{b^\rho - a^\rho}{\rho} \right)^\alpha \right) \right.$$
$$\left. + \frac{K}{(1-M)} \left(\frac{|\vartheta_1|}{\Gamma(1+\alpha)} + \frac{\Gamma(\gamma)}{\Gamma(\gamma+\alpha)} \right) \left(\frac{b^\rho - a^\rho}{\rho} \right)^\alpha \right] \|u - w\|_{PC_{\gamma,\rho}}.$$

By (4.19), the operator Ψ is a contraction. Hence, by Theorem 2.45, Ψ has a unique fixed point $u^* \in PC_{\gamma,\rho}(J)$.

Step 2: We show that such a fixed point $u^* \in PC_{\gamma,\rho}(J)$ is actually in $PC_{\gamma,\rho}^\gamma(J)$. Since u^* is the unique fixed point of operator Ψ in $PC_{\gamma,\rho}(J)$, then for each $t \in J_k$, with $k = 0, \dots, m$, we have

$$u^*(t) = \frac{1}{\Gamma(\gamma)} \left(\frac{t^\rho - t_k^\rho}{\rho} \right)^{\gamma-1} \left[\vartheta_2 - \vartheta_1 \sum_{i=1}^m L_i(u(t_i^-)) - \vartheta_1 \sum_{i=1}^m \left({}^\rho \mathcal{J}_{(t_{i-1})^+}^{1-\gamma+\alpha} h \right)(t_i) \right.$$
$$\left. - \vartheta_1 \left({}^\rho \mathcal{J}_{t_m^+}^{1-\gamma+\alpha} h \right)(b) + \sum_{a < t_k < t} L_k(u(t_k^-)) + \sum_{a < t_k < t} \left({}^\rho \mathcal{J}_{(t_{k-1})^+}^{1-\gamma+\alpha} h \right)(t_k) \right]$$
$$+ \left({}^\rho \mathcal{J}_{t_k^+}^\alpha h \right)(t),$$

where $h \in PC_{\gamma,\rho}(J)$ such that

$$h(t) = f(t, u^*(t), h(t)).$$

Applying ${}^\rho \mathcal{D}_{t_k^+}^\gamma$ to both sides and by Lemmas 2.19 and 2.33, we have

$$\begin{aligned}
{}^\rho \mathcal{D}_{t_k^+}^\gamma u^*(t) &= \left({}^\rho \mathcal{D}_{t_k^+}^\gamma \, {}^\rho \mathcal{J}_{t_k^+}^\alpha f(s, u^*(s), h(s)) \right)(t) \\
&= \left({}^\rho \mathcal{D}_{t_k^+}^{\beta(1-\alpha)} f(s, u^*(s), h(s)) \right)(t).
\end{aligned}$$

Since $\gamma \geq \alpha$, by (4.3.1), the right-hand side is in $PC_{\gamma,\rho}(J)$ and thus $^{\rho}\mathcal{D}^{\gamma}_{t_k^+}u^* \in PC_{\gamma,\rho}(J)$ which implies that $u^* \in PC^{\gamma}_{\gamma,\rho}(J)$. As a consequence of Steps 1 and 2 together with Theorem 4.3, we can conclude that the problem (4.1)–(4.3) has a unique solution in $PC^{\gamma}_{\gamma,\rho}(J)$.   □

Our second result is based on Schaefer's fixed point theorem.

Theorem 4.4 *Assume that in addition of the hypothesis (4.3.1)–(4.3.3), the following hold.*

(4.4.1) There exist functions p_1, p_2, $p_3 \in C([a,b], \mathbb{R}_+)$ with

$$p_1^* = \sup_{t\in[a,b]} p_1(t), \quad p_2^* = \sup_{t\in[a,b]} p_2(t), \quad p_3^* = \sup_{t\in[a,b]} p_3(t) < 1$$

such that

$$|f(t, u, w)| \leq p_1(t) + p_2(t)|u| + p_3(t)|w| \ \text{for } t \in J \text{ and } u, w \in \mathbb{R}.$$

(4.4.2) The functions $L_k : \mathbb{R} \longrightarrow \mathbb{R}$ are continuous and there exist constants Φ_1, $\Phi_2 > 0$ such that

$$|L_k(u)| \leq \Phi_1|u| + \Phi_2 \ \text{for each } u \in \mathbb{R}, k = 1, \ldots, m.$$

If

$$(|\vartheta_1| + 1)\left(\frac{m\Phi_1 p^*}{\Gamma(\gamma)} + \frac{mp_2^*(b^{\rho}-a^{\rho})^{\alpha}}{(1-p_3^*)\Gamma(1+\alpha)\rho^{\alpha}}\right) + \left(\frac{|\vartheta_1|}{\Gamma(1+\alpha)} + \frac{\Gamma(\gamma)}{\Gamma(\gamma+\alpha)}\right)\left(\frac{p_2^*(b^{\rho}-a^{\rho})^{\alpha}}{(1-p_3^*)\rho^{\alpha}}\right) < 1, \quad (4.20)$$

then the problem (4.1)–(4.3) has at least one solution in $PC^{\gamma}_{\gamma,\rho}(J)$.

Proof We shall use Schaefer's fixed point theorem to prove in several steps that the operator Ψ defined in (4.18) has a fixed point.

Step 1: Ψ is continuous.
Let $\{u_n\}$ be a sequence such that $u_n \to u$ in $PC_{\gamma,\rho}(J)$.
Then for each $t \in J$, we have

$$\left| ((\Psi u_n)(t) - (\Psi u)(t))\left(\frac{t^{\rho} - t_k^{\rho}}{\rho}\right)^{1-\gamma} \right|$$

$$\leq \frac{1}{\Gamma(\gamma)}\left[|\vartheta_1|\sum_{i=1}^{m}|L_i(u_n(t_i^-)) - L_i(u(t_i^-))| + |\vartheta_1|\left(^{\rho}\mathcal{J}^{1-\gamma+\alpha}_{t_m^+}|h_n(s) - h(s)|\right)\right] \quad (b)$$

$$+ |\vartheta_1| \sum_{i=1}^{m} \left({}^{\rho}\mathcal{J}_{(t_{i-1})^+}^{1-\gamma+\alpha} |h_n(s) - h(s)| \right)(t_i) + \sum_{a < t_k < t} |L_k(u_n(t_k^-)) - L_k(u(t_k^-))|$$

$$+ \sum_{a < t_k < t} \left({}^{\rho}\mathcal{J}_{(t_{k-1})^+}^{1-\gamma+\alpha} |h_n(s) - h(s)| \right)(t_k) \Bigg]$$

$$+ \left(\frac{t^{\rho} - t_k^{\rho}}{\rho} \right)^{1-\gamma} \left({}^{\rho}\mathcal{J}_{t_k^+}^{\alpha} |h_n(s) - h(s)| \right)(t),$$

where $h_n, h \in PC_{\gamma,\rho}(J)$ such that

$$h_n(t) = f(t, u_n(t), h_n(t)),$$
$$h(t) = f(t, u(t), h(t)).$$

Since $u_n \to u$, then we get $h_n(t) \to h(t)$ as $n \to \infty$ for each $t \in J$, and since f and L_k are continuous, then we have

$$\|\Psi u_n - \Psi u\|_{PC_{\gamma,\rho}} \to 0 \text{ as } n \to \infty.$$

Step 2: We show that Ψ is the mapping of two bounded sets in $PC_{\gamma,\rho}(J)$.
For $\eta > 0$, there exists a positive constant β such that $B_\eta = \{u \in PC_{\gamma,\rho}(J) : \|u\|_{PC_{\gamma,\rho}} \leq \eta\}$, we have $\|\Psi(u)\|_{PC_{\gamma,\rho}} \leq \beta$.
By (4.4.1) and from (4.18), we have for each $t \in J_k, k = 0, \ldots, m,$

$$\left| \left(\frac{t^{\rho} - t_k^{\rho}}{\rho} \right)^{1-\gamma} h(t) \right| = \left| \left(\frac{t^{\rho} - t_k^{\rho}}{\rho} \right)^{1-\gamma} f(t, u(t), h(t)) \right|$$

$$\leq \left(\frac{t^{\rho} - t_k^{\rho}}{\rho} \right)^{1-\gamma} (p_1(t) + p_2(t)|u(t)| + p_3(t)|h(t)|),$$

which implies that

$$\|h\|_{PC_{\gamma,\rho}} \leq p_1^* \left(\frac{b^{\rho} - a^{\rho}}{\rho} \right)^{1-\gamma} + p_2^* \eta + p_3^* \|h\|_{PC_{\gamma,\rho}}.$$

Then

$$\|h\|_{PC_{\gamma,\rho}} \leq \frac{p_1^* \left(\frac{b^{\rho} - a^{\rho}}{\rho} \right)^{1-\gamma} + p_2^* \eta}{1 - p_3^*} := \Lambda.$$

Thus (4.18) implies

$$\left| \left(\frac{t^\rho - t_k^\rho}{\rho} \right)^{1-\gamma} (\Psi u)(t) \right|$$

$$\leq \frac{1}{\Gamma(\gamma)} \left[|\vartheta_2| + |\vartheta_1| \sum_{i=1}^{m} |L_i(u(t_i^-))| + |\vartheta_1| \sum_{i=1}^{m} \left({}^\rho \mathcal{J}_{(t_{i-1})^+}^{1-\gamma+\alpha} |h(s)| \right)(t_i) \right.$$

$$+ |\vartheta_1| \left({}^\rho \mathcal{J}_{t_m^+}^{1-\gamma+\alpha} |h(s)| \right)(b) + \sum_{a < t_k < t} |L_k(u(t_k^-))|$$

$$\left. + \sum_{a < t_k < t} \left({}^\rho \mathcal{J}_{(t_{k-1})^+}^{1-\gamma+\alpha} |h(s)| \right)(t_k) \right] + \left(\frac{t^\rho - t_k^\rho}{\rho} \right)^{1-\gamma} \left({}^\rho \mathcal{J}_{t_k^+}^{\alpha} |h(s)| \right)(t).$$

Then

$$\left| \left(\frac{t^\rho - t_k^\rho}{\rho} \right)^{1-\gamma} (\Psi u)(t) \right|$$

$$\leq \left[|\vartheta_2| + |\vartheta_1| m(\Phi_1 p^* \eta + \Phi_2) + |\vartheta_1| m \Lambda \left({}^\rho \mathcal{J}_{(t_{k-1})^+}^{1-\gamma+\alpha} \left(\frac{s^\rho - t_{k-1}^\rho}{\rho} \right)^{\gamma-1} \right)(t_k) \right.$$

$$+ |\vartheta_1| \Lambda \left({}^\rho \mathcal{J}_{t_m^+}^{1-\gamma+\alpha} \left(\frac{s^\rho - t_m^\rho}{\rho} \right)^{\gamma-1} \right)(b) + m(\Phi_1 p^* \eta + \Phi_2)$$

$$\left. + m \Lambda \left({}^\rho \mathcal{J}_{(t_{k-1})^+}^{1-\gamma+\alpha} \left(\frac{s^\rho - t_{k-1}^\rho}{\rho} \right)^{\gamma-1} \right)(t_k) \right] \frac{1}{\Gamma(\gamma)}$$

$$+ \Lambda \left(\frac{t^\rho - t_k^\rho}{\rho} \right)^{1-\gamma} \left({}^\rho \mathcal{J}_{t_k^+}^{\alpha} \left(\frac{s^\rho - t_k^\rho}{\rho} \right)^{\gamma-1} \right)(t).$$

By Lemma 2.19, we have

$$\|\Psi u\|_{PC_{\gamma,\rho}} \leq (|\vartheta_1| + 1) \left(\frac{m(\Phi_1 p^* \eta + \Phi_2)}{\Gamma(\gamma)} + \frac{m\Lambda}{\Gamma(1+\alpha)} \left(\frac{b^\rho - a^\rho}{\rho} \right)^\alpha \right)$$

$$+ \Lambda \left(\frac{|\vartheta_1|}{\Gamma(1+\alpha)} + \frac{\Gamma(\gamma)}{\Gamma(\gamma+\alpha)} \right) \left(\frac{b^\rho - a^\rho}{\rho} \right)^\alpha + \frac{|\vartheta_2|}{\Gamma(\gamma)}$$

$$:= \beta.$$

Step 3: Ψ maps bounded sets into equicontinuous sets of $PC_{\gamma,\rho}$.

Let $\epsilon_1, \epsilon_2 \in J, \epsilon_1 < \epsilon_2, B_\eta$ be a bounded set of $PC_{\gamma,\rho}$ as in Step 2, and let $u \in B_\eta$. Then

$$\left| \left(\frac{\epsilon_1^\rho - t_k^\rho}{\rho} \right)^{1-\gamma} (\Psi u)(\epsilon_1) - \left(\frac{\epsilon_2^\rho - t_k^\rho}{\rho} \right)^{1-\gamma} (\Psi u)(\epsilon_2) \right|$$

$$\leq \frac{1}{\Gamma(\gamma)} \left[\sum_{\epsilon_1 < t_k < \epsilon_2} |L_k(u(t_k^-))| + \sum_{\epsilon_1 < t_k < \epsilon_2} \left({}^\rho \mathcal{J}_{(t_{k-1})^+}^{1-\gamma+\alpha} |h(s)| \right)(t_k) \right]$$

$$+ \frac{\Lambda \Gamma(\gamma)}{\Gamma(\gamma+\alpha)} \left| \left(\frac{\epsilon_1^\rho - t_k^\rho}{\rho} \right)^\alpha - \left(\frac{\epsilon_2^\rho - t_k^\rho}{\rho} \right)^\alpha \right|.$$

As $\epsilon_1 \to \epsilon_2$, the right-hand side of the above inequality tends to zero. From steps 1 to 3 with the Arzela-Ascoli theorem, we conclude that $\Psi : PC_{\gamma,\rho} \to PC_{\gamma,\rho}$ is continuous and completely continuous.

Step 4: A priori bound. Now it remains to show that the set

$$G = \{ u \in PC_{\gamma,\rho} : u = \lambda^* \Psi(u) \text{ for some } 0 < \lambda^* < 1 \}$$

is bounded. Let $u \in G$, then $u = \lambda^* \Psi(u)$ for some $0 < \lambda^* < 1$.
By (4.4.1), we have for each $t \in J$,

$$\left| \left(\frac{t^\rho - t_k^\rho}{\rho} \right)^{1-\gamma} h(t) \right| = \left| \left(\frac{t^\rho - t_k^\rho}{\rho} \right)^{1-\gamma} f(t, u(t), h(t)) \right|$$

$$\leq \left(\frac{t^\rho - t_k^\rho}{\rho} \right)^{1-\gamma} (p_1(t) + p_2(t)|u(t)| + p_3(t)|h(t)|),$$

which implies that

$$\|h\|_{PC_{\gamma,\rho}} \leq p_1^* \left(\frac{b^\rho - a^\rho}{\rho} \right)^{1-\gamma} + p_2^* \|u\|_{PC_{\gamma,\rho}} + p_3^* \|h\|_{PC_{\gamma,\rho}},$$

then

$$\|h\|_{PC_{\gamma,\rho}} \leq \frac{p_1^* \left(\frac{b^\rho - a^\rho}{\rho} \right)^{1-\gamma} + p_2^* \|u\|_{PC_{\gamma,\rho}}}{1 - p_3^*}.$$

This implies, by (4.18), (4.4.2) and by letting the estimation of Step 2, that for each $t \in J$ we have

$$\|u\|_{PC_{\gamma,\rho}} \le (|\vartheta_1| + 1)\left(\frac{m(\Phi_1 p^*\|u\|_{PC_{\gamma,\rho}} + \Phi_2)}{\Gamma(\gamma)} + \frac{1}{(1 - p_3^*)\Gamma(1 + \alpha)}\right.$$

$$\left. \times \left[mp_1^*\left(\frac{b^\rho - a^\rho}{\rho}\right)^{1+\alpha-\gamma} + mp_2^*\|u\|_{PC_{\gamma,\rho}}\left(\frac{b^\rho - a^\rho}{\rho}\right)^\alpha\right]\right)$$

$$+ \frac{p_1^*\left(\frac{b^\rho - a^\rho}{\rho}\right)^{1+\alpha-\gamma} + p_2^*\|u\|_{PC_{\gamma,\rho}}\left(\frac{b^\rho - a^\rho}{\rho}\right)^\alpha}{(1 - p_3^*)}$$

$$\times \left(\frac{|\vartheta_1|}{\Gamma(1 + \alpha)} + \frac{\Gamma(\gamma)}{\Gamma(\gamma + \alpha)}\right) + \frac{|\vartheta_2|}{\Gamma(\gamma)},$$

$$\le \left[(|\vartheta_1| + 1)\left(\frac{m\Phi_1 p^*}{\Gamma(\gamma)} + \frac{mp_2^*(b^\rho - a^\rho)^\alpha}{(1 - p_3^*)\Gamma(1 + \alpha)\rho^\alpha}\right)\right.$$

$$\left. + \left(\frac{|\vartheta_1|}{\Gamma(1 + \alpha)} + \frac{\Gamma(\gamma)}{\Gamma(\gamma + \alpha)}\right)\left(\frac{p_2^*(b^\rho - a^\rho)^\alpha}{(1 - p_3^*)\rho^\alpha}\right)\right]\|u\|_{PC_{\gamma,\rho}}$$

$$+ \frac{|\vartheta_2|}{\Gamma(\gamma)} + (|\vartheta_1| + 1)\left(\frac{m\Phi_2}{\Gamma(\gamma)} + \frac{mp_1^*(b^\rho - a^\rho)^{1+\alpha-\gamma}}{(1 - p_3^*)\Gamma(1 + \alpha)\rho^{1+\alpha-\gamma}}\right)$$

$$+ \left(\frac{|\vartheta_1|}{\Gamma(1 + \alpha)} + \frac{\Gamma(\gamma)}{\Gamma(\gamma + \alpha)}\right)\left(\frac{p_1^*(b^\rho - a^\rho)^{1+\alpha-\gamma}}{(1 - p_3^*)\rho^{1+\alpha-\gamma}}\right).$$

By (4.20), we have

$$\|u\|_{PC_{\gamma,\rho}}$$

$$\le \frac{\frac{|\vartheta_2|}{\Gamma(\gamma)} + \left[(|\vartheta_1|+1)\left(\frac{m\Phi_2}{\Gamma(\gamma)} + \frac{mp_1^*(b^\rho-a^\rho)^{1+\alpha-\gamma}}{(1-p_3^*)\Gamma(1+\alpha)\rho^{1+\alpha-\gamma}}\right) + \left(\frac{|\vartheta_1|}{\Gamma(1+\alpha)} + \frac{\Gamma(\gamma)}{\Gamma(\gamma+\alpha)}\right)\left(\frac{p_1^*(b^\rho-a^\rho)^{1+\alpha-\gamma}}{(1-p_3^*)\rho^{1+\alpha-\gamma}}\right)\right]}{1 - \left[(|\vartheta_1|+1)\left(\frac{m\Phi_1 p^*}{\Gamma(\gamma)} + \frac{mp_2^*(b^\rho-a^\rho)^\alpha}{(1-p_3^*)\Gamma(1+\alpha)\rho^\alpha}\right) + \left(\frac{|\vartheta_1|}{\Gamma(1+\alpha)} + \frac{\Gamma(\gamma)}{\Gamma(\gamma+\alpha)}\right)\left(\frac{p_2^*(b^\rho-a^\rho)^\alpha}{(1-p_3^*)\rho^\alpha}\right)\right]}$$

$$:= R.$$

As a consequence of Theorem 2.47, and using Step 2 of the last result, we deduce that Ψ has a fixed point which is a solution of the problem (4.1)–(4.3). □

Our third result is based on Krasnoselskii's fixed point theorem.

Theorem 4.5 *Assume that* (4.3.1), (4.4.1), *and* (4.4.2) *hold. If*

$$(|\vartheta_1| + 1)\left(\frac{m\Phi_1 p^*}{\Gamma(\gamma)} + \frac{mp_2^*(b^\rho - a^\rho)^\alpha}{(1 - p_3^*)\Gamma(1 + \alpha)\rho^\alpha}\right) + \frac{p_2^*|\vartheta_1|(b^\rho - a^\rho)^\alpha}{(1 - p_3^*)\Gamma(1 + \alpha)\rho^\alpha} < 1, \quad (4.21)$$

then the problem (4.1)–(4.3) *has at least one solution in* $PC_{\gamma,\rho}^\gamma(J)$.

Proof Consider the set

$$B_\eta = \{u \in PC_{\gamma,\rho}(J) : ||u||_{PC_{\gamma,\rho}} \le \eta\},$$

where

$$\eta \ge \frac{(|\vartheta_1|+1)\left(\frac{m\Phi_2}{\Gamma(\gamma)} + \frac{m\Lambda}{\Gamma(1+\alpha)}\left(\frac{b^\rho - a^\rho}{\rho}\right)^\alpha\right) + \Lambda\left(\frac{|\vartheta_1|}{\Gamma(1+\alpha)} + \frac{\Gamma(\gamma)}{\Gamma(\gamma+\alpha)}\right)\left(\frac{b^\rho - a^\rho}{\rho}\right)^\alpha + \frac{|\vartheta_2|}{\Gamma(\gamma)}}{1 - (|\vartheta_1|+1)\frac{m\Phi_1 p^*}{\Gamma(\gamma)}}.$$

We define the operators Q_1 and Q_2 on B_η by

$$Q_1 u(t) = \frac{1}{\Gamma(\gamma)}\left(\frac{t^\rho - t_k^\rho}{\rho}\right)^{\gamma-1}\left[\vartheta_2 - \vartheta_1 \sum_{i=1}^{m} L_i(u(t_i^-)) - \vartheta_1 \sum_{i=1}^{m}\left({}^\rho\mathcal{J}_{(t_{i-1})^+}^{1-\gamma+\alpha}h\right)(t_i)\right.$$
$$\left. - \vartheta_1 \left({}^\rho\mathcal{J}_{t_m^+}^{1-\gamma+\alpha}h\right)(b) + \sum_{a<t_k<t} L_k(u(t_k^-)) + \sum_{a<t_k<t}\left({}^\rho\mathcal{J}_{(t_{k-1})^+}^{1-\gamma+\alpha}h\right)(t_k)\right],$$

$$\tag{4.22}$$

$$Q_2 u(t) = \left({}^\rho\mathcal{J}_{t_k^+}^\alpha h\right)(t), \tag{4.23}$$

where $k = 0, \ldots, m$ and $h : J \to \mathbb{R}$ be a function satisfying the functional equation

$$h(t) = f(t, u(t), h(t)).$$

Then the fractional integral equation (4.18) can be written as operator equation

$$\Psi u(t) = Q_1 u(t) + Q_2 u(t), \quad u \in PC_{\gamma,\rho}(J).$$

The proof will be given in several steps.

Step 1: We prove that $Q_1 u + Q_2 w \in B_\eta$ for any $u, z \in B_\eta$.
Same as Step 2 of the last result, by (4.4.1), (4.4.2), and Lemma 2.19, for each $t \in J$, we have

$$||Q_1 u + Q_2 w||_{PC_{\gamma,\rho}} \le ||Q_1 u||_{PC_{\gamma,\rho}} + ||Q_2 w||_{PC_{\gamma,\rho}}$$
$$\le (|\vartheta_1| + 1)\left(\frac{m(\Phi_1 p^* \eta + \Phi_2)}{\Gamma(\gamma)} + \frac{m\Lambda}{\Gamma(1+\alpha)}\left(\frac{b^\rho - a^\rho}{\rho}\right)^\alpha\right)$$
$$+ \Lambda\left(\frac{|\vartheta_1|}{\Gamma(1+\alpha)} + \frac{\Gamma(\gamma)}{\Gamma(\gamma+\alpha)}\right)\left(\frac{b^\rho - a^\rho}{\rho}\right)^\alpha + \frac{|\vartheta_2|}{\Gamma(\gamma)}.$$

Since

$$\eta \geq \frac{(|\vartheta_1|+1)\left(\frac{m\Phi_2}{\Gamma(\gamma)}+\frac{m\Lambda}{\Gamma(1+\alpha)}\left(\frac{b^\rho-a^\rho}{\rho}\right)^\alpha\right)+\Lambda\left(\frac{|\vartheta_1|}{\Gamma(1+\alpha)}+\frac{\Gamma(\gamma)}{\Gamma(\gamma+\alpha)}\right)\left(\frac{b^\rho-a^\rho}{\rho}\right)^\alpha+\frac{|\vartheta_2|}{\Gamma(\gamma)}}{1-(|\vartheta_1|+1)\frac{m\Phi_1 p^*}{\Gamma(\gamma)}},$$

we have

$$\|Q_1 y + Q_2 z\|_{PC_{\gamma,\rho}} \leq \eta,$$

which infers that $Q_1 u + Q_2 w \in B_\eta$.

Step 2: Q_1 is a contraction.
Let $u, w \in PC_{\gamma,\rho}(J)$ and $t \in J$.
By (4.4.1), we have

$$|h(t) - g(t)| = |f(t, u(t), h(t)) - f(t, w(t), g(t))|$$
$$\leq p_2(t)|u(t) - w(t)| + p_3(t)|h(t) - g(t)|.$$

Then,

$$|h(t) - g(t)| \leq \frac{p_2(t)}{1 - p_3(t)}|u(t) - w(t)| \leq \frac{p_2^*}{1 - p_3^*}|u(t) - w(t)|,$$

where $p_1^* = \sup_{t \in [a,b]} p_1(t)$, $p_2^* = \sup_{t \in [a,b]} p_2(t)$ and $h, g \in C([a, b], \mathbb{R})$ such that

$$h(t) = f(t, u(t), h(t)),$$
$$g(t) = f(t, w(t), g(t)).$$

Then by (4.4.2) and using the estimation in Step 1 of the first result, we have

$$|Q_1 y(t) - Q_1 z(t)|$$
$$\leq \frac{1}{\Gamma(\gamma)}\left(\frac{t^\rho - t_k^\rho}{\rho}\right)^{\gamma-1}\|u - w\|_{PC_{\gamma,\rho}}\left[|\vartheta_1|m\Phi_1 p^*\right.$$
$$+\frac{p_2^*|\vartheta_1|\Gamma(\gamma)}{(1-p_3^*)\Gamma(1+\alpha)}\left(\frac{b^\rho-t_m^\rho}{\rho}\right)^\alpha+\frac{mp_2^*|\vartheta_1|\Gamma(\gamma)}{(1-p_3^*)\Gamma(1+\alpha)}\left(\frac{t_k^\rho-t_{k-1}^\rho}{\rho}\right)^\alpha$$
$$\left.+m\Phi_1 p^* + \frac{mp_2^*\Gamma(\gamma)}{(1-p_3^*)\Gamma(1+\alpha)}\left(\frac{t_k^\rho-t_{k-1}^\rho}{\rho}\right)^\alpha\right],$$

hence

$$\|Q_1 u - Q_1 w\|_{PC_{\gamma,\rho}} \leq \left[(|\vartheta_1|+1)\left(\frac{m\Phi_1 p^*}{\Gamma(\gamma)}+\frac{mp_2^*}{(1-p_3^*)\Gamma(1+\alpha)}\left(\frac{b^\rho-a^\rho}{\rho}\right)^\alpha\right)\right.$$
$$\left.+\frac{p_2^*|\vartheta_1|}{(1-p_3^*)\Gamma(1+\alpha)}\left(\frac{b^\rho-a^\rho}{\rho}\right)^\alpha\right]\|u - w\|_{PC_{\gamma,\rho}}.$$

By (4.21), the operator Q_1 is a contraction.

Step 3: Q_2 is continuous and compact.

The continuity of Q_2 follows from the continuity of f. Next we prove that Q_2 is uniformly bounded on B_η. Let any $w \in B_\eta$. By using the estimation in Step 2 of the last result, (4.23) implies

$$\left| \left(\frac{t^\rho - t_k^\rho}{\rho} \right)^{1-\gamma} (Q_2 z)(t) \right| \leq \left(\frac{t^\rho - t_k^\rho}{\rho} \right)^{1-\gamma} \left({}^\rho \mathcal{J}_{t_k^+}^\alpha |g(s)| \right)(t),$$

$$\leq \Lambda \left(\frac{t^\rho - t_k^\rho}{\rho} \right)^{1-\gamma} \left({}^\rho \mathcal{J}_{t_k^+}^\alpha \left(\frac{s^\rho - t_k^\rho}{\rho} \right)^{\gamma-1} \right)(t),$$

where $k = 0, \dots, m$ and $g : J \to \mathbb{R}$ be a function satisfying the functional equation

$$g(t) = f(t, w(t), g(t)).$$

By Lemma 2.19, we have

$$\| Q_2 z \|_{PC_{\gamma,\rho}} \leq \frac{\Lambda \Gamma(\gamma)}{\Gamma(\gamma + \alpha)} \left(\frac{b^\rho - a^\rho}{\rho} \right)^\alpha.$$

This means that Q_2 is uniformly bounded on B_η. Next, we show that $Q_2 B_\eta$ is equicontinuous. Let any $w \in B_\eta$ and $a < \epsilon_1 < \epsilon_2 \leq b$. Then

$$\left| \left(\frac{\epsilon_1^\rho - t_k^\rho}{\rho} \right)^{1-\gamma} (Q_2 z)(\epsilon_1) - \left(\frac{\epsilon_2^\rho - t_k^\rho}{\rho} \right)^{1-\gamma} (Q_2 z)(\epsilon_2) \right|$$

$$\leq \frac{\Lambda \Gamma(\gamma)}{\Gamma(\gamma + \alpha)} \left| \left(\frac{\epsilon_1^\rho - t_k^\rho}{\rho} \right)^\alpha - \left(\frac{\epsilon_2^\rho - t_k^\rho}{\rho} \right)^\alpha \right|.$$

Note that

$$\left| \left(\frac{\epsilon_1^\rho - t_k^\rho}{\rho} \right)^{1-\gamma} (Q_2 z)(\epsilon_1) - \left(\frac{\epsilon_2^\rho - t_k^\rho}{\rho} \right)^{1-\gamma} (Q_2 z)(\epsilon_2) \right| \to 0 \quad \text{as} \quad \epsilon_1 \to \epsilon_2.$$

This shows that $Q_2 B_\eta$ is equicontinuous on J. Therefore, $Q_2 B_\eta$ is relatively compact. By PC_γ-type Arzela-Ascoli Theorem, Q_2 is compact.

As a consequence of Theorem 2.50, we deduce that Ψ has at least a fixed point $u^* \in PC_{\gamma,\rho}(J)$, and by the same way of the proof of Theorem 4.3, we can easily show that $u^* \in PC_{\gamma,\rho}^\gamma(J)$. Using Lemma 4.2, we conclude that the problem (4.1)–(4.3) has at least one solution in the space $PC_{\gamma,\rho}^\gamma(J)$. □

4.2.2 Ulam-Hyers-Rassias Stability

Now we are concerned with the Ulam-Hyers-Rassias Stability of our problem (4.1)–(4.3). Let $u \in PC_{\gamma,\rho}(J)$, $\epsilon > 0$, $\tau > 0$, and $\theta : J \longrightarrow [0, \infty)$ be a continuous function. We consider the following inequality:

$$\begin{cases} \left| \left({}^{\rho}\mathcal{D}^{\alpha,\beta}_{t^+_k} u \right)(t) - f\left(t, u(t), \left({}^{\rho}\mathcal{D}^{\alpha,\beta}_{t^+_k} u \right)(t) \right) \right| \le \epsilon\theta(t),\, t \in J_k,\, k = 0, \dots, m, \\[3mm] \left| \left({}^{\rho}\mathcal{J}^{1-\gamma}_{t^+_k} u \right)(t^+_k) - \left({}^{\rho}\mathcal{J}^{1-\gamma}_{t^+_{k-1}} u \right)(t^-_k) - L_k(u(t^-_k)) \right| \le \epsilon\tau,\, k = 1, \dots, m. \end{cases}$$

$$(4.24)$$

Definition 1 ([156]) Problem (4.1)–(4.3) is Ulam-Hyers-Rassias (U-H-R) stable with respect to (θ, τ) if there exists a real number $a_{f,m,\theta} > 0$ such that for each $\epsilon > 0$ and for each solution $u \in PC_{\gamma,\rho}(J)$ of inequality (4.24) there exists a solution $w \in PC_{\gamma,\rho}(J)$ of (4.1)–(4.3) with

$$|u(t) - w(t)| \le \epsilon a_{f,m,\theta}(\theta(t) + \tau), \quad t \in (a, b].$$

Remark 4.6 ([156]) A function $u \in PC_{\gamma,\rho}(J)$ is a solution of inequality (4.24) if and only if there exist $\sigma \in PC_{\gamma,\rho}(J)$ and a sequence $\sigma_k, k = 0, \dots, m$ such that

1. $|\sigma(t)| \le \epsilon\theta(t)$ and $|\sigma_k| \le \epsilon\tau, t \in J_k, k = 1, \dots, m$;
2. $\left({}^{\rho}\mathcal{D}^{\alpha,\beta}_{t^+_k} u \right)(t) = f\left(t, u(t), \left({}^{\rho}\mathcal{D}^{\alpha,\beta}_{t^+_k} u \right)(t) \right) + \sigma(t), t \in J_k, k = 0, \dots, m$;
3. $\left({}^{\rho}\mathcal{J}^{1-\gamma}_{t^+_k} u \right)(t^+_k) = \left({}^{\rho}\mathcal{J}^{1-\gamma}_{t^+_{k-1}} u \right)(t^-_k) + L_k(u(t^-_k)) + \sigma_k, k = 1, \dots, m$.

Theorem 4.7 *Assume that in addition to (4.3.1)–(4.3.3) and (4.19), the following hypothesis holds:*

(4.8.1) *There exist a nondecreasing function $\theta \in PC_{\gamma,\rho}(J)$ and $\lambda_\theta, \tilde{\lambda}_\theta > 0$ such that for each $t \in (a, b]$, we have*
$$({}^{\rho}\mathcal{J}^{\alpha}_{a^+}\theta)(t) \le \lambda_\theta\theta(t),$$

and
$$({}^{\rho}\mathcal{J}^{1-\gamma}_{a^+}\theta)(t) \le \tilde{\lambda}_\theta\theta(t).$$

Then Eq. (4.1) is U-H-R stable with respect to (θ, τ).

Proof Consider the operator Ψ defined in (4.18). Let $u \in PC_{\gamma,\rho}(J)$ be a solution of inequality (4.24), and let us assume that w is the unique solution of the problem

$$\begin{cases} \left({}^{\rho}\mathcal{D}_{t_k^+}^{\alpha,\beta} w \right)(t) = f\left(t, w(t), \left({}^{\rho}\mathcal{D}_{t_k^+}^{\alpha,\beta} w \right)(t) \right); \ t \in J_k, \ k = 0, \ldots, m, \\[2mm] \left({}^{\rho}\mathcal{J}_{t_k^+}^{1-\gamma} w \right)(t_k^+) = \left({}^{\rho}\mathcal{J}_{t_{k-1}^+}^{1-\gamma} w \right)(t_k^-) + L_k(w(t_k^-)); \ k = 1, \ldots, m, \\[2mm] c_1 \left({}^{\rho}\mathcal{J}_{a^+}^{1-\gamma} w \right)(a^+) + c_2 \left({}^{\rho}\mathcal{J}_{t_m^+}^{1-\gamma} w \right)(b) = c_3, \\[2mm] \left({}^{\rho}\mathcal{J}_{a^+}^{1-\gamma} w \right)(a^+) = \left({}^{\rho}\mathcal{J}_{a^+}^{1-\gamma} u \right)(a^+). \end{cases}$$

By Lemma 4.2, we obtain for each $t \in J$

$$w(t) = \left[\left({}^{\rho}\mathcal{J}_{a^+}^{1-\gamma} w \right)(a^+) + \sum_{a<t_k<t} L_k(w(t_k^-)) + \sum_{a<t_k<t} \left({}^{\rho}\mathcal{J}_{(t_{k-1})^+}^{1-\gamma+\alpha} h \right)(t_k) \right]$$

$$\times \frac{1}{\Gamma(\gamma)} \left(\frac{t^{\rho} - t_k^{\rho}}{\rho} \right)^{\gamma-1} + \left({}^{\rho}\mathcal{J}_{t_k^+}^{\alpha} h \right)(t), \ t \in J_k, k = 0, \ldots, m,$$

where $h : J \to \mathbb{R}$ be a function satisfying the functional equation

$$h(t) = f(t, w(t), h(t)).$$

Since u is a solution of the inequality (4.24), by Remark 4.6, we have

$$\begin{cases} \left({}^{\rho}\mathcal{D}_{t_k^+}^{\alpha,\beta} u \right)(t) = f\left(t, u(t), \left({}^{\rho}\mathcal{D}_{t_k^+}^{\alpha,\beta} u \right)(t) \right) + \sigma(t), t \in J_k, k = 0, \ldots, m; \\[2mm] \left({}^{\rho}\mathcal{J}_{t_k^+}^{1-\gamma} u \right)(t_k^+) = \left({}^{\rho}\mathcal{J}_{t_{k-1}^+}^{1-\gamma} u \right)(t_k^-) + L_k(u(t_k^-)) + \sigma_k, k = 1, \ldots, m. \end{cases} \quad (4.25)$$

Clearly, the solution of (4.25) is given by

$$u(t) = \frac{1}{\Gamma(\gamma)} \left(\frac{t^{\rho} - t_k^{\rho}}{\rho} \right)^{\gamma-1} \left[\left({}^{\rho}\mathcal{J}_{a^+}^{1-\gamma} u \right)(a^+) + \sum_{a<t_k<t} L_k(u(t_k^-)) + \sum_{a<t_k<t} \sigma_k \right.$$

$$+ \sum_{a<t_k<t} \left({}^{\rho}\mathcal{J}_{(t_{k-1})^+}^{1-\gamma+\alpha} g \right)(t_k) + \sum_{a<t_k<t} \left({}^{\rho}\mathcal{J}_{(t_{k-1})^+}^{1-\gamma+\alpha} \sigma \right)(t_k) \Big]$$

$$+ \left({}^{\rho}\mathcal{J}_{t_k^+}^{\alpha} g \right)(t) + \left({}^{\rho}\mathcal{J}_{t_k^+}^{\alpha} \sigma \right)(t) \quad t \in J_k, k = 0, \ldots, m,$$

where $g : J \to \mathbb{R}$ be a function satisfying the functional equation

$$g(t) = f(t, u(t), g(t)).$$

Hence, for each $t \in J$, we have

$$|u(t) - w(t)| \le \left[\sum_{k=1}^{m} |L_k(u(t_k^-)) - L_k(w(t_k^-))| + \sum_{k=1}^{m} \left({}^{\rho}\mathcal{J}_{(t_{k-1})^+}^{1-\gamma+\alpha} |\sigma(s)| \right)(t_k) \right.$$

$$\left. + \sum_{k=1}^{m} \left({}^{\rho}\mathcal{J}_{(t_{k-1})^+}^{1-\gamma+\alpha} |g(s) - h(s)| \right)(t_k) + \sum_{k=1}^{m} |\sigma_k| \right] \frac{\left(\frac{t^{\rho} - t_k^{\rho}}{\rho} \right)^{\gamma-1}}{\Gamma(\gamma)}$$

$$+ \left({}^{\rho}\mathcal{J}_{t_k^+}^{\alpha} |g(s) - h(s)| \right)(t) + \left({}^{\rho}\mathcal{J}_{t_k^+}^{\alpha} |\sigma(s)| \right)(t).$$

Thus,

$$\|u - w\|_{PC_{\gamma,\rho}} \le \frac{1}{\Gamma(\gamma)} \left[m\epsilon\tau + (m\tilde{\lambda}_{\theta} + 1)\epsilon\lambda_{\theta}\theta(t) + \sum_{k=1}^{m} l^* |u(t_k^-) - w(t_k^-)| \right.$$

$$\left. + \sum_{k=1}^{m} \left({}^{\rho}\mathcal{J}_{(t_{k-1})^+}^{1-\gamma+\alpha} |g(s) - h(s)| \right)(t_k) \right]$$

$$+ \left(\frac{t^{\rho} - t_k^{\rho}}{\rho} \right)^{1-\gamma} \left({}^{\rho}\mathcal{J}_{t_k^+}^{\alpha} |g(s) - h(s)| \right)(t).$$

By condition (4.3.2) and Lemma 2.19, for $t \in J$, we have

$$\|u - w\|_{PC_{\gamma,\rho}} \le \frac{1}{\Gamma(\gamma)} \left[m\epsilon\tau + (m\tilde{\lambda}_{\theta} + 1)\epsilon\lambda_{\theta}\theta(t) + ml^* p^* \|u - w\|_{PC_{\gamma,\rho}} \right]$$

$$+ \left[\frac{mK}{(1 - M)\Gamma(1 + \alpha)} \left(\frac{t_k^{\rho} - t_{k-1}^{\rho}}{\rho} \right)^{\alpha} \right.$$

$$\left. + \frac{K\Gamma(\gamma)}{(1 - M)\Gamma(\gamma + \alpha)} \left(\frac{t^{\rho} - t_k^{\rho}}{\rho} \right)^{\alpha} \right] \|u - w\|_{PC_{\gamma,\rho}}.$$

Thus,

$$\|u - w\|_{PC_{\gamma,\rho}} \le \frac{1}{\Gamma(\gamma)} \left(m\epsilon\tau + (m\tilde{\lambda}_{\theta} + 1)\epsilon\lambda_{\theta}\theta(t) \right)$$

$$+ \left[\frac{ml^* p^*}{\Gamma(\gamma)} + \frac{K}{1 - M} \left(\frac{m}{\Gamma(1 + \alpha)} + \frac{\Gamma(\gamma)}{\Gamma(\gamma + \alpha)} \right) \left(\frac{b^{\rho} - a^{\rho}}{\rho} \right)^{\alpha} \right]$$

$$\times \|u - w\|_{PC_{\gamma,\rho}}.$$

Then by (4.19), we have

$$\|u - w\|_{PC_{\gamma,\rho}} \le a_{\theta}\epsilon(\tau + \theta(t)),$$

where

$$a_\theta = \frac{1}{\Gamma(\gamma)} (m + (m\tilde{\lambda}_\theta + 1)\lambda_\theta) \left[1 - \frac{ml^* p^*}{\Gamma(\gamma)} \right.$$

$$\left. + \frac{K}{1-M} \left(\frac{m}{\Gamma(1+\alpha)} + \frac{\Gamma(\gamma)}{\Gamma(\gamma+\alpha)} \right) \left(\frac{b^\rho - a^\rho}{\rho} \right)^\alpha \right]^{-1}.$$

Hence, Eq. (4.1) is U-H-R stable with respect to (θ, τ).   □

4.2.3   Examples

Example 4.8 Consider the following impulsive boundary value problem of generalized Hilfer fractional differential equation

$$\left(\tfrac{1}{2} \mathcal{D}_{t_k^+}^{\frac{1}{2},0} u \right)(t) = \frac{1}{97e^{t+2} \left(1 + |u(t)| + \left| \tfrac{1}{2} \mathcal{D}_{t_k^+}^{\frac{1}{2},0} u(t) \right| \right)} + \frac{\ln(e + \sqrt{t})}{e^2 \sqrt{t-1}}, \ t \in J_0 \cup J_1,$$

(4.26)

$$\left(\tfrac{1}{2} \mathcal{J}_{e^+}^{\frac{1}{2}} u \right)(e^+) - \left(\tfrac{1}{2} \mathcal{J}_{1^+}^{\frac{1}{2}} u \right)(e^-) = \frac{|u(e^-)|}{3 + |u(e^-)|},$$

(4.27)

$$3 \left(\tfrac{1}{2} \mathcal{J}_{1^+}^{\frac{1}{2}} u \right)(1^+) - 2 \left(\tfrac{1}{2} \mathcal{J}_{e^+}^{\frac{1}{2}} u \right)(3) = 0,$$

(4.28)

where $J_0 = (1, e]$, $J_1 = (e, 3]$, $t_0 = 1$, and $t_1 = e$.
Set

$$f(t, u, w) = \frac{1}{97e^{t+2}(1 + |u| + |w|)} + \frac{\ln(e + \sqrt{t})}{e^2 \sqrt{t-1}}, \ t \in (1, 3], \ u, w \in \mathbb{R}.$$

We have

$$PC_{\gamma,\rho}^{\beta(1-\alpha)}([1, 3]) = PC_{\frac{1}{2},\frac{1}{2}}^0([1, 3])$$

$$= \left\{ g : (1, 3] \to \mathbb{R} : \sqrt{2} \left(\sqrt{t} - \sqrt{t_k} \right)^{\frac{1}{2}} g \in PC([1, 3]) \right\},$$

with $\gamma = \alpha = \frac{1}{2}$, $\rho = \frac{1}{2}$, $\beta = 0$,, and $k \in \{0, 1\}$. Clearly, the continuous function $f \in PC_{\frac{1}{2},\frac{1}{2}}^0([1, 3])$. Hence, the condition (4.3.1) is satisfied.
For each $u, \bar{u}, w, \bar{w} \in \mathbb{R}$ and $t \in (1, 3]$:

$$|f(t, u, w) - f(t, \bar{u}, \bar{w})| \le \frac{1}{97e^{t+2}} (|u - \bar{u}| + |w - \bar{w}|)$$

$$\le \frac{1}{97e^3} (|u - \bar{u}| + |w - \bar{w}|).$$

Hence, condition (4.3.2) is satisfied with $K = M = \frac{1}{97e^3}$.
And let

$$L_1(u) = \frac{u}{3+u}, \quad u \in [0, \infty).$$

Let $u, w \in [0, \infty)$. Then we have

$$|L_1(u) - L_1(w)| = \left|\frac{u}{3+u} - \frac{w}{3+w}\right| = \frac{3|u - w|}{(3+u)(3+w)} \leq \frac{1}{3}|u - w|,$$

and so the condition (4.3.3) is satisfied and $l^* = \frac{1}{3}$.
A simple computation shows that the condition (4.19) of Theorem 4.3 is satisfied for

$$L = \frac{1}{\sqrt{2\pi}(\sqrt{e}-1)} + \frac{3\sqrt{2}(\sqrt{3}-1)^{\frac{1}{2}}}{(97e^3-1)\Gamma(\frac{3}{2})} + \frac{\sqrt{2}(\sqrt{3}-1)^{\frac{1}{2}}}{(97e^3-1)}\left(\frac{2}{\Gamma(\frac{3}{2})} + \sqrt{\pi}\right)$$
$$\approx 0.52720987569 < 1.$$

Then the problem (4.26)–(4.28) has a unique solution in $PC^{\frac{1}{2}}_{\frac{1}{2},\frac{1}{2}}([1, 3])$.
Also, hypothesis (4.8.1) is satisfied with

$$\theta(t) = e^5, \tau = 1 \text{ and } \lambda_\theta = \tilde{\lambda}_\theta = \frac{2}{\Gamma(\frac{3}{2})}.$$

Indeed, for each $t \in J_0 \cup J_1$, we get

$$({}^\rho \mathcal{J}^{\frac{1}{2}}_{1^+}\theta)(t) \leq \frac{2e^5}{\Gamma(\frac{3}{2})}$$
$$= \lambda_\theta \theta(t) = \tilde{\lambda}_\theta \theta(t).$$

Consequently, Theorem 4.7 implies that Eq. (4.26) is U-H-R stable.

Example 4.9   Consider the following impulsive initial value problem of generalized Hilfer fractional differential equation

$$\left({}^1 D^{\frac{1}{2},0}_{t^+_k} u\right)(t) = \frac{3 + |u(t)| + |{}^1 D^{\frac{1}{2},0}_{t^+_k} u(t)|}{53e^{-t+4}(1 + |u(t)| + |{}^1 D^{\frac{1}{2},0}_{t^+_k} u(t)|)}, \quad \text{for each } t \in J_0 \cup J_1, \quad (4.29)$$

$$\left({}^1 \mathcal{J}^{\frac{1}{2}}_{e^+} u\right)(e^+) - \left({}^1 \mathcal{J}^{\frac{1}{2}}_{1^+} u\right)(e^-) = \frac{|u(e^-)|}{2 + |u(e^-)|}, \quad (4.30)$$

$$\left({}^1 \mathcal{J}^{\frac{1}{2}}_{1^+} u\right)(1^+) = 0, \quad (4.31)$$

where $J_0 = (1, e]$, $J_1 = (e, 3]$, $t_0 = 1$ and $t_1 = e$.

Set
$$f(t, u, w) = \frac{3 + |u| + |w|}{53e^{-t+4}(1 + |u| + |w|)}, \quad t \in (1, 3], \; u, w \in \mathbb{R}.$$

We have
$$PC_{\gamma, \rho}^{\beta(1-\alpha)}([1, 3]) = PC_{\frac{1}{2}, 1}^0([1, 3]) = \left\{ g : (1, 3] \to \mathbb{R} : (\sqrt{t - t_k})g \in PC([1, 3]) \right\},$$

with $\gamma = \alpha = \frac{1}{2}$, $\rho = 1$, $\beta = 0$ and $k \in \{0, 1\}$.
Clearly, the continuous function $f \in PC_{\frac{1}{2}, 1}^0([1, 3])$. Hence the condition (4.3.1) is satisfied.
For each $u, w \in \mathbb{R}$ and $t \in (1, 3]$:
$$|f(t, u, w)| \le \frac{1}{53e^{-t+4}}(3 + |u| + |w|).$$

Hence, condition (4.4.1) is satisfied with
$$p_1(t) = \tfrac{3}{53e^{-t+4}}, \; p_2(t) = p_3(t) = \tfrac{1}{53e^{-t+4}},$$

and
$$p_1^* = \tfrac{3}{53e}, \; p_2^* = p_3^* = \tfrac{1}{53e}.$$

And let
$$L_1(u) = \frac{u}{2 + u}, \; u \in [0, \infty).$$

Then we have
$$|L_1(u)| \le \frac{1}{2}|u| + 2,$$

and so the condition (4.4.2) is satisfied with $\Phi_1 = \frac{1}{2}$ and $\Phi_2 = 2$.
The condition (4.20) of Theorem 4.4 is satisfied for
$$(|\vartheta_1| + 1)\left(\frac{m\Phi_1 p^*}{\Gamma(\gamma)} + \frac{mp_2^*(b^\rho - a^\rho)^\alpha}{(1 - p_3^*)\Gamma(1+\alpha)\rho^\alpha} \right) + \left(\frac{|\vartheta_1|}{\Gamma(1+\alpha)} + \frac{\Gamma(\gamma)}{\Gamma(\gamma+\alpha)} \right)\left(\frac{p_2^*(b^\rho - a^\rho)^\alpha}{(1 - p_3^*)\rho^\alpha} \right)$$
$$= \left(\frac{1}{2\sqrt{2\pi}} + \frac{\sqrt{2}}{(53e - 1)\Gamma(\frac{3}{2})} \right) + \frac{\sqrt{2\pi}}{53e - 1}$$
$$\approx 0.22814541069 \le 1.$$

Then the problem (4.29)–(4.31) has at least one solution in $PC_{\frac{1}{2}, 1}^{\frac{1}{2}}([1, 3])$. Also, hypothesis (4.8.1) is satisfied with
$$\theta(t) = t - 1, \; \tau = 1 \text{ and } \lambda_\theta = \tilde{\lambda}_\theta = \frac{\sqrt{2}\Gamma(2)}{\Gamma(\frac{5}{2})}.$$

Indeed, for each $t \in J_0 \cup J_1$, we get

$$(^P\mathcal{J}_{1^+}^{\frac{1}{2}}\theta)(t) \le \frac{\sqrt{2}\Gamma(2)}{\Gamma(\frac{5}{2})}(t-1)$$
$$= \lambda_\theta\theta(t) = \tilde{\lambda}_\theta\theta(t).$$

Consequently, by a simple change of the constants l^*, K, and M from hypothesis (4.3.1) and (4.3.2) to Φ_1, p_2^*, and p_3^* from (4.4.1) and (4.4.2), Theorem 4.7 implies that Eq. (4.29) is G.U-H-R stable.

Example 4.10 Consider the following impulsive anti-periodic boundary value problem of generalized Hilfer fractional differential equation:

$$\left(^1\mathcal{D}_{t_k^+}^{\frac{1}{2},0}u\right)(t) = \frac{e^2 + |u(t)| + |^1\mathcal{D}_{t_k^+}^{\frac{1}{2},0}u(t)|}{77e^{-t+2}(1+|u(t)|+|^1\mathcal{D}_{t_k^+}^{\frac{1}{2},0}u(t)|)}, \quad t \in J_k; k=0,\ldots,4, \quad (4.32)$$

$$\left(^1\mathcal{J}_{t_k^+}^{\frac{1}{2}}u\right)(t_k^+) - \left(^1\mathcal{J}_{t_{(k-1)}^+}^{\frac{1}{2}}u\right)(t_k^-) = \frac{|u(t_k^-)|}{10k+|u(t_k^-)|}; k=1,\ldots,4, \quad (4.33)$$

$$\left(^1\mathcal{J}_{1^+}^{\frac{1}{2}}u\right)(1^+) = -\left(^1\mathcal{J}_{9^+}^{\frac{1}{2}}u\right)(2), \quad (4.34)$$

where $J_k = (t_k, t_{k+1}]$, $t_k = 1 + \frac{k}{5}$ for $k=0,\ldots,4$, $m=4$, $a=t_0=1$, and $b=t_5=2$.
Set
$$f(t,u,w) = \frac{e^2+|u|+|w|}{77e^{-t+2}(1+|u|+|w|)}, \quad t \in (1,2], u,w \in \mathbb{R}.$$

We have
$$PC_{\gamma,\rho}^{\beta(1-\alpha)}([1,2]) = PC_{\frac{1}{2},1}^0([1,2]) = \{g:(1,2]\to\mathbb{R}:(\sqrt{t-t_k})g \in PC([1,2])\},$$

with $\gamma=\alpha=\frac{1}{2}$, $\rho=1$, $\beta=0$, and $k=0,\ldots,4$.
Clearly, the continuous function $f \in PC_{\frac{1}{2},1}^0([1,2])$. So, the condition (4.3.1) is satisfied.
For each $u,w \in \mathbb{R}$ and $t \in (1,2]$:

$$|f(t,u,w)| \le \frac{1}{77e^{-t+2}}(e^2+|u|+|w|).$$

Hence, the condition (4.4.1) is satisfied with

$$p_1(t) = \frac{e^2}{77e^{-t+2}}, \; p_2(t) = p_3(t) = \frac{1}{77e^{-t+2}},$$

and

$$p_1^* = \frac{e^2}{77}, \ p_2^* = p_3^* = \frac{1}{77}.$$

And let

$$L_k(u) = \frac{u}{10k + u}, \ k = 1, \ldots, 4, \ u \in [0, \infty).$$

Then we have

$$|L_k(u)| \le \frac{1}{10}|u| + 1, \ k = 1, \ldots, 4,$$

and so the condition (4.4.2) is satisfied with $\Phi_1 = \frac{1}{10}$ and $\Phi_2 = 1$.
The condition (4.21) of Theorem 4.5 is satisfied for

$$(|\vartheta_1| + 1) \left(\frac{m\Phi_1 p^*}{\Gamma(\gamma)} + \frac{mp_2^*(b^\rho - a^\rho)^\alpha}{(1-p_3^*)\Gamma(1+\alpha)\rho^\alpha} \right) + \frac{p_2^*|\vartheta_1|(b^\rho - a^\rho)^\alpha}{(1-p_3^*)\Gamma(1+\alpha)\rho^\alpha} = \frac{3\sqrt{5}}{5\sqrt{\pi}} + \frac{125}{1463\Gamma(\frac{3}{2})} < 1.$$

Then the problem (4.32)–(4.34) has at least one solution in $PC_{\frac{1}{2},1}^{\frac{1}{2}}([1, 2])$. Also, hypothesis
(4.8.1) is satisfied with

$$\theta(t) = (1 - t)^2, \ \tau = 1 \text{ and } \lambda_\theta = \tilde{\lambda}_\theta = \frac{\Gamma(3)}{\Gamma(\frac{7}{2})}.$$

Indeed, for each $t \in J_k, k = 0, \ldots, 4$, we get

$$({}^\rho \mathcal{J}_{1^+}^{\frac{1}{2}} \theta)(t) \le \frac{\Gamma(3)}{\Gamma(\frac{7}{2})}(t - 1)^2$$
$$= \lambda_\theta \theta(t) = \tilde{\lambda}_\theta \theta(t).$$

Same as Example 4.9, Theorem 4.7 implies that Eq. (4.32) is U-H-R stable.

4.3   Existence and Ulam Stability Results for Generalized Hilfer-Type Boundary Value Problem

Motivated by the works mentioned in the introduction of the current chapter, in this section, we discuss the existence results to the boundary value problem with nonlinear implicit generalized Hilfer-type fractional differential equation with instantaneous impulses:

$$\left({}^\rho \mathcal{D}_{t_k^+}^{\alpha, \beta} u \right)(t) = f\left(t, u(t), \left({}^\rho \mathcal{D}_{t_k^+}^{\alpha, \beta} u \right)(t) \right); \ t \in J_k, \ k = 0, \cdots, m, \quad (4.35)$$

$$\left({}^\rho \mathcal{J}_{t_k^+}^{1-\gamma} u \right)(t_k^+) = \left({}^\rho \mathcal{J}_{t_{k-1}^+}^{1-\gamma} u \right)(t_k^-) + \varpi_k(u(t_k^-)); \ k = 1, \cdots, m, \quad (4.36)$$

$$c_1 \left({}^\rho \mathcal{J}_{a^+}^{1-\gamma} u \right)(a^+) + c_2 \left({}^\rho \mathcal{J}_{t_m^+}^{1-\gamma} u \right)(b) = c_3, \quad (4.37)$$

where $^{\rho}\mathcal{D}_{t_k^+}^{\alpha,\beta}$, $^{\rho}\mathcal{J}_{t_k^+}^{1-\gamma}$ are the generalized Hilfer fractional derivative of order $\alpha \in (0,1)$ and type $\beta \in [0,1]$ and generalized Hilfer fractional integral of order $1 - \gamma$, $(\gamma = \alpha + \beta - \alpha\beta)$, respectively, c_1, c_2 are reals with $c_1 + c_2 \neq 0$, $J_k := (t_k, t_{k+1}]; k = 0, \cdots, m$, $a = t_0 < t_1 < \cdots < t_m < t_{m+1} = b < \infty$, $u(t_k^+) = \lim_{\epsilon \to 0^+} u(t_k + \epsilon)$ and $u(t_k^-) = \lim_{\epsilon \to 0^-} u(t_k + \epsilon)$ represent the right- and left-hand limits of $u(t)$ at $t = t_k$, $c_3 \in E$, $f : J \times E \times E \to E$ is a given function, and $\varpi_k : E \to E; k = 1, \cdots, m$ are given continuous functions.

4.3.1 Existence Results

Consider the weighted Banach space

$$PC_{\gamma,\rho}(J) = \Big\{ u : J \to E : u(t) \in C(J_k, E); k = 0, \cdots, m, \text{ and there exist}$$

$$u(t_k^-) \text{ and } \left(^{\rho}\mathcal{J}_{t_k^+}^{1-\gamma} u \right)(t_k^+); k = 0, \cdots, m, \text{ with } u(t_k^-) = u(t_k) \Big\},$$

and

$$PC_{\gamma,\rho}^n(J) = \Big\{ u \in PC^{n-1} : u^{(n)} \in PC_{\gamma,\rho}(J) \Big\}, n \in \mathbb{N},$$

$$PC_{\gamma,\rho}^0(J) = PC_{\gamma,\rho}(J),$$

with the norm

$$\|u\|_{PC_{\gamma,\rho}} = \max_{k=0,\ldots,m} \left\{ \sup_{t \in [t_k, t_{k+1}]} \left\| \left(\frac{t^\rho - t_k^\rho}{\rho} \right)^{1-\gamma} u(t) \right\| \right\}.$$

We define the space

$$PC_{\gamma,\rho}^\gamma(J) = \Big\{ u \in PC_{\gamma,\rho}(J), \ ^{\rho}\mathcal{D}_{t_k^+}^{\gamma} u \in PC_{\gamma,\rho}(J) \Big\}, \ k = 0, \ldots, m.$$

Lemma 4.11 ([75]) *Let $D \subset PC_{\gamma,\rho}(J)$ be a bounded and equicontinuous set, then*
(i) the function $t \to \mu(D(t))$ is continuous on J, and

$$\mu_{PC_{\gamma,\rho}}(D) = \sup_{t \in [a,b]} \mu \left(\left(\frac{t^\rho - t_k^\rho}{\rho} \right)^{1-\gamma} D(t) \right),$$

(ii) $\mu \left(\left\{ \int_a^b u(s)ds : u \in D \right\} \right) \le \int_a^b \mu(D(s))ds$, *where*

$$D(t) = \{u(t) : t \in D\}, t \in J.$$

By following the same results from the previous section, we have the following result.

Lemma 4.12 *Let* $\gamma = \alpha + \beta - \alpha\beta$ *where* $0 < \alpha < 1$ *and* $0 \le \beta \le 1$, *let* $f : J \times E \times E \to E$ *be a function such that* $f(\cdot, u(\cdot), w(\cdot)) \in PC_{\gamma,\rho}(J)$ *for any* $u, w \in PC_{\gamma,\rho}(J)$. *If* $u \in PC_{\gamma,\rho}^{\gamma}(J)$, *then* u *satisfies the problem (4.35)–(4.37) if and only if* u *is the fixed point of the operator* $\Psi : PC_{\gamma,\rho}(J) \to PC_{\gamma,\rho}(J)$ *defined by*

$$\Psi u(t) = \frac{1}{\Gamma(\gamma)} \left(\frac{t^\rho - t_k^\rho}{\rho} \right)^{\gamma-1} \left[\vartheta_2 - \vartheta_1 \sum_{i=1}^m \varpi_i(u(t_i^-)) - \vartheta_1 \sum_{i=1}^m \left({}^\rho J_{(t_{i-1})^+}^{1-\gamma+\alpha} h \right)(t_i) \right.$$
$$\left. - \vartheta_1 \left({}^\rho J_{t_m^+}^{1-\gamma+\alpha} h \right)(b) + \sum_{a < t_k < t} \varpi_k(u(t_k^-)) + \sum_{a < t_k < t} \left({}^\rho J_{(t_{k-1})^+}^{1-\gamma+\alpha} h \right)(t_k) \right]$$
$$+ \left({}^\rho J_{t_k^+}^{\alpha} h \right)(t) \qquad t \in J_k, k = 0, \cdots, m, \tag{4.38}$$

where $h : J \to \mathbb{R}$ *be a function satisfying the functional equation:*

$$h(t) = f(t, u(t), h(t)).$$

We are now in a position to state and prove our existence result for the problem (4.35)–(4.37) based on Mönch's fixed point theorem.

Theorem 4.13 *Assume that the hypotheses that follow are met.*

(4.14.1) The function $t \mapsto f(t, u, w)$ *is measurable and continuous on* J *for each* $u, w \in E$, *and the functions* $u \mapsto f(t, u, w)$ *and* $w \mapsto f(t, u, w)$ *are continuous on* E *for a.e.* $t \in J$, *and*

$$f(\cdot, u(\cdot), w(\cdot)) \in PC_{\gamma,\rho}^{\beta(1-\alpha)} \text{ for any } u, w \in PC_{\gamma,\rho}(J).$$

(4.14.2) There exists a continuous function $p : [a, b] \longrightarrow [0, \infty)$ *such that*

$$\|f(t, u, w)\| \le p(t), \text{ for a.e. } t \in J \text{ and for each } u, w \in E.$$

(4.14.3) For each bounded set $B \subset E$ *and for each* $t \in (a, b]$, *we have*

$$\mu(f(t, B, ({}^\rho D_{a^+}^{\alpha,\beta} B))) \le \left(\frac{t^\rho - t_k^\rho}{\rho} \right)^{1-\gamma} p(t)\mu(B),$$

where $^{\rho}D_{a+}^{\alpha,\beta}B = \{^{\rho}D_{a+}^{\alpha,\beta}w : w \in B\}$ and $k = 1, \cdots, m$.

(4.14.4) *The functions* $\varpi_k : E \longrightarrow E$ *are continuous and there exists* $\eta^* > 0$ *such that*

$$\|\varpi_k(u)\| \le \eta^*\|u\| \text{ for each } u \in E, k = 1, \cdots, m.$$

(4.14.5) *For each bounded set* $B \subset E$ *and for each* $t \in J$, *we have*

$$\mu(\varpi_k(B)) \le \eta^* \left(\frac{t^{\rho} - t_k^{\rho}}{\rho}\right)^{1-\gamma} \mu(B), k = 1, \cdots, m.$$

If

$$\mathfrak{L} := \frac{m\eta^*}{\Gamma(\gamma)} + p^* \left(\frac{1}{\Gamma(\alpha+1)} + \frac{m}{\Gamma(\gamma)\Gamma(2-\gamma+\alpha)}\right) \left(\frac{b^{\rho}-a^{\rho}}{\rho}\right)^{1-\gamma+\alpha} < 1, \tag{4.39}$$

where $p^* = \sup\limits_{t \in [a,b]} p(t)$, *then the problem* (4.35)–(4.37) *has at least one solution in* $PC_{\gamma,\rho}^{\gamma}(J)$.

Proof Consider the operator $\Psi : PC_{\gamma,\rho}(J) \to PC_{\gamma,\rho}(J)$ defined in (4.38) and the ball $B_R := B(0, R) = \{w \in PC_{\gamma,\rho}(J) : \|w\|_{PC_{\gamma,\rho}} \le R\}$.
For any $u \in B_R$, and each $t \in J$ we have

$$\left\|\left(\frac{t^{\rho}-t_k^{\rho}}{\rho}\right)^{1-\gamma}(\Psi u)(t)\right\|$$

$$\le \frac{1}{\Gamma(\gamma)}\left[\|\vartheta_2\| + |\vartheta_1|\sum_{i=1}^{m}\|\varpi_i(u(t_i^-))\| + |\vartheta_1|\sum_{i=1}^{m}\left(^{\rho}\mathcal{J}_{(t_{i-1})^+}^{1-\gamma+\alpha}\|h(s)\|\right)(t_i)\right.$$

$$+|\vartheta_1|\left(^{\rho}\mathcal{J}_{t_m^+}^{1-\gamma+\alpha}\|h\|\right)(b) + \sum_{a<t_k<t}\|\varpi_k(u(t_k^-))\|$$

$$\left.+ \sum_{a<t_k<t}\left(^{\rho}\mathcal{J}_{(t_{k-1})^+}^{1-\gamma+\alpha}\|h(s)\|\right)(t_k)\right] + \left(\frac{t^{\rho}-t_k^{\rho}}{\rho}\right)^{1-\gamma}\left(^{\rho}\mathcal{J}_{t_k^+}^{\alpha}\|h(s)\|\right)(t)$$

$$\le \frac{\|\vartheta_2\|}{\Gamma(\gamma)} + \frac{|\vartheta_1|+1}{\Gamma(\gamma)}\left(ml^* R + mp^*\left(^{\rho}\mathcal{J}_{(t_{i-1})^+}^{1-\gamma+\alpha}(1)\right)(t_i)\right)$$

$$+ \frac{|\vartheta_1|p^*}{\Gamma(\gamma)}\left(^{\rho}\mathcal{J}_{t_m^+}^{1-\gamma+\alpha}(1)\right)(b) + p^*\left(\frac{t^{\rho}-t_k^{\rho}}{\rho}\right)^{1-\gamma}\left(^{\rho}\mathcal{J}_{t_k^+}^{\alpha}(1)\right)(t).$$

By Lemma 2.19, we have

$$\left\|\left(\frac{t^{\rho}-t_k^{\rho}}{\rho}\right)^{1-\gamma}(\Psi u)(t)\right\|$$

$$\le \frac{\|\vartheta_2\|}{\Gamma(\gamma)} + \frac{|\vartheta_1|+1}{\Gamma(\gamma)}\left(ml^* R + \frac{mp^*}{\Gamma(2-\gamma+\alpha)}\left(\frac{t_i^{\rho}-t_{i-1}^{\rho}}{\rho}\right)^{1-\gamma+\alpha}\right)$$

$$+ \frac{|\vartheta_1|p^*}{\Gamma(\gamma)\Gamma(2-\gamma+\alpha)}\left(\frac{b^{\rho}-t_m^{\rho}}{\rho}\right)^{1-\gamma+\alpha} + \frac{p^*}{\Gamma(\alpha+1)}\left(\frac{t^{\rho}-t_k^{\rho}}{\rho}\right)^{1-\gamma+\alpha}.$$

Hence, for any $u \in PC_{\gamma,\rho}(J)$, and each $t \in (a, b]$ we get

$$\|(\Psi u)\|_{PC_{\gamma,\rho}} \leq \frac{\|\vartheta_2\|}{\Gamma(\gamma)} + \frac{|\vartheta_1|+1}{\Gamma(\gamma)} \left[ml^* R + \frac{mp^*}{\Gamma(2-\gamma+\alpha)} \left(\frac{b^\rho - a^\rho}{\rho} \right)^{1-\gamma+\alpha} \right]$$
$$+ \left(\frac{|\vartheta_1| p^*}{\Gamma(\gamma)\Gamma(2-\gamma+\alpha)} + \frac{p^*}{\Gamma(\alpha+1)} \right) \left(\frac{b^\rho - a^\rho}{\rho} \right)^{1-\gamma+\alpha}$$
$$\leq R.$$

This proves that Ψ transforms the ball B_R into itself. We shall show that the operator $\Psi : B_R \rightarrow B_R$ satisfies all the assumptions of Theorem 2.49. The proof will be given in several steps.

Step 1: $\Psi : B_R \rightarrow B_R$ is continuous.
Let $\{u_n\}$ be a sequence such that $u_n \rightarrow u$ in $PC_{\gamma,\rho}(J)$.
Then for each $t \in (a, b]$, we have

$$\left\| ((\Psi u_n)(t) - (\Psi u)(t)) \left(\frac{t^\rho - t_k^\rho}{\rho} \right)^{1-\gamma} \right\|$$

$$\leq \left[|\vartheta_1| \sum_{i=1}^{m} \|\varpi_i(u_n(t_i^-)) - \varpi_i(u(t_i^-))\| + |\vartheta_1| \left({}^\rho \mathcal{J}_{t_m^+}^{1-\gamma+\alpha} \|h_n(s) - h(s)\| \right) \right. \quad (b)$$

$$+ |\vartheta_1| \sum_{i=1}^{m} \left({}^\rho \mathcal{J}_{(t_{i-1})^+}^{1-\gamma+\alpha} \|h_n(s) - h(s)\| \right)(t_i) + \sum_{a < t_k < t} \|\varpi_k(u_n(t_k^-)) - \varpi_k(u(t_k^-))\|$$

$$+ \sum_{a < t_k < t} \left({}^\rho \mathcal{J}_{(t_{k-1})^+}^{1-\gamma+\alpha} \|h_n(s) - h(s)\| \right)(t_k) \left. \right] \frac{1}{\Gamma(\gamma)}$$

$$+ \left(\frac{t^\rho - t_k^\rho}{\rho} \right)^{1-\gamma} \left({}^\rho \mathcal{J}_{t_k^+}^{\alpha} \|h_n(s) - h(s)\| \right)(t),$$

where $h_n, h \in PC_{\gamma,\rho}$ such that

$$h_n(t) = f(t, u_n(t), h_n(t)),$$
$$h(t) = f(t, u(t), h(t)).$$

Since $u_n \rightarrow u$, then we get $h_n(t) \rightarrow h(t)$ as $n \rightarrow \infty$ for each $t \in J$, and by the Lebesgue dominated convergence theorem, we have

$$\|\Psi u_n - \Psi u\|_{PC_{\gamma,\rho}} \rightarrow 0 \text{ as } n \rightarrow \infty.$$

Step 2: $\Psi(B_R)$ is bounded and equicontinuous.
Since $\Psi(B_R) \subset B_R$ and B_R is bounded, then $\Psi(B_R)$ is bounded.
Next, let $\epsilon_1, \epsilon_2 \in J$, $\epsilon_1 < \epsilon_2$, and let $u \in B_R$. Then

$$\left\| \left(\tfrac{\epsilon_1^\rho - t_k^\rho}{\rho}\right)^{1-\gamma} (\Psi u)(\epsilon_1) - \left(\tfrac{\epsilon_2^\rho - t_k^\rho}{\rho}\right)^{1-\gamma} (\Psi u)(\epsilon_2) \right\|$$

$$\leq \tfrac{1}{\Gamma(\gamma)} \left[\sum_{\epsilon_1 < t_k < \epsilon_2} \|\varpi_k(u(t_k^-))\| + \sum_{\epsilon_1 < t_k < \epsilon_2} \left(^\rho \mathcal{J}_{(t_{k-1})^+}^{1-\gamma+\alpha} \|h(s)\|\right)(t_k) \right]$$

$$+ \tfrac{p^*}{\Gamma(\alpha+1)} \left| \left(\tfrac{\epsilon_1^\rho - t_k^\rho}{\rho}\right)^{1-\gamma+\alpha} - \left(\tfrac{\epsilon_2^\rho - t_k^\rho}{\rho}\right)^{1-\gamma+\alpha} \right|.$$

As $\epsilon_1 \to \epsilon_2$, the right-hand side of the above inequality tends to zero. Hence, $\Psi(B_R)$ is bounded and equicontinuous.

Step 3: The implication (2.11) of Theorem 2.49 holds.

Now let D be an equicontinuous subset of B_R such that $D \subset \overline{\Psi(D)} \cup \{0\}$; therefore, the function $t \longrightarrow d(t) = \mu(D(t))$ is continuous on J. By (4.14.3), (4.14.5), and the properties of the measure μ, for each $t \in J$, we have

$$\left(\frac{t^\rho - t_k^\rho}{\rho}\right)^{1-\gamma} d(t) \leq \mu \left(\left(\frac{t^\rho - t_k^\rho}{\rho}\right)^{1-\gamma} (\Psi D)(t) \cup \{0\} \right)$$

$$\leq \mu \left(\left(\frac{t^\rho - t_k^\rho}{\rho}\right)^{1-\gamma} (\Psi D)(t) \right)$$

$$\leq \frac{1}{\Gamma(\gamma)} \left[\sum_{a < t_k < t} \eta^* \left(\frac{t^\rho - t_k^\rho}{\rho}\right)^{1-\gamma} \mu(D(t)) \right.$$

$$+ \sum_{a < t_k < t} \left(^\rho \mathcal{J}_{(t_{k-1})^+}^{1-\gamma+\alpha} \left(\frac{s^\rho - t_k^\rho}{\rho}\right)^{1-\gamma} p(s)\mu(D(s))\right)(t_k) \right]$$

$$+ \left(\frac{t^\rho - t_k^\rho}{\rho}\right)^{1-\gamma} \left(^\rho \mathcal{J}_{t_k^+}^\alpha \left(\frac{s^\rho - t_k^\rho}{\rho}\right)^{1-\gamma} p(s)\mu(D(s))\right)(t)$$

$$\leq p^* \left(\frac{b^\rho - a^\rho}{\rho}\right)^{1-\gamma} \left(^\rho \mathcal{J}_{a+}^\alpha \left(\frac{s^\rho - t_k^\rho}{\rho}\right)^{1-\gamma} d(s)\right)(t)$$

$$+ \frac{m\eta^* \|d\|_{PC_{\gamma,\rho}}}{\Gamma(\gamma)} + \frac{mp^*}{\Gamma(\gamma)} \left[^\rho \mathcal{J}_{a+}^{1-\gamma+\alpha} \left(\frac{s^\rho - t_k^\rho}{\rho}\right)^{1-\gamma} d(s)\right](t)$$

$$\leq \left[\frac{m\eta^*}{\Gamma(\gamma)} + \frac{p^*}{\Gamma(\alpha+1)} \left(\frac{b^\rho - a^\rho}{\rho}\right)^{1-\gamma+\alpha} \right.$$

$$+ \frac{mp^*}{\Gamma(\gamma)\Gamma(2-\gamma+\alpha)} \left(\frac{b^\rho - a^\rho}{\rho}\right)^{1-\gamma+\alpha} \right] \|d\|_{PC_{\gamma,\rho}}.$$

Thus

$$\|d\|_{PC_{\gamma,\rho}} \leq \mathfrak{L} \|d\|_{PC_{\gamma,\rho}}.$$

From (4.39), we get $\|d\|_{PC_{\gamma,\rho}} = 0$, that is $d(t) = \mu(D(t)) = 0$, for each $t \in J_k, k = 0, \cdots, m$, and then $D(t)$ is relatively compact in E. In view of the Ascoli-Arzela theorem, D is relatively compact in B_R. Applying now Theorem 2.49, we conclude that Ψ has a fixed point $u^* \in PC_{\gamma,\rho}(J)$, which is solution of the problem (4.35)–(4.37).

Step 4: We show that such a fixed point $u^* \in PC_{\gamma,\rho}(J)$ is actually in $PC_{\gamma,\rho}^{\gamma}(J)$. Since u^* is the unique fixed point of operator Ψ in $PC_{\gamma,\rho}(J)$, then for each $t \in J_k$, with $k = 0, \cdots, m$, we have

$$u^*(t) = \frac{1}{\Gamma(\gamma)} \left(\frac{t^\rho - t_k^\rho}{\rho} \right)^{\gamma-1} \left[\vartheta_2 - \vartheta_1 \sum_{i=1}^{m} \varpi_i(u(t_i^-)) - \vartheta_1 \sum_{i=1}^{m} \left({}^\rho \mathcal{J}_{(t_{i-1})^+}^{1-\gamma+\alpha} h \right)(t_i) \right.$$
$$\left. - \vartheta_1 \left({}^\rho \mathcal{J}_{t_m^+}^{1-\gamma+\alpha} h \right)(b) + \sum_{a < t_k < t} \varpi_k(u(t_k^-)) + \sum_{a < t_k < t} \left({}^\rho \mathcal{J}_{(t_{k-1})^+}^{1-\gamma+\alpha} h \right)(t_k) \right]$$
$$+ \left({}^\rho \mathcal{J}_{t_k^+}^{\alpha} h \right)(t),$$

where $h \in PC_{\gamma,\rho}(J)$ such that

$$h(t) = f(t, u^*(t), h(t)).$$

Applying ${}^\rho \mathcal{D}_{t_k^+}^{\gamma}$ to both sides and by Lemmas 2.19 and 2.33, we have

$$ {}^\rho \mathcal{D}_{t_k^+}^{\gamma} u^*(t) = \left({}^\rho \mathcal{D}_{t_k^+}^{\gamma} \, {}^\rho \mathcal{J}_{t_k^+}^{\alpha} f(s, u^*(s), h(s)) \right)(t) $$
$$ = \left({}^\rho \mathcal{D}_{t_k^+}^{\beta(1-\alpha)} f(s, u^*(s), h(s)) \right)(t). $$

Since $\gamma \geq \alpha$, by (4.14.1), the right-hand side is in $PC_{\gamma,\rho}(J)$ and thus ${}^\rho \mathcal{D}_{t_k^+}^{\gamma} u^* \in PC_{\gamma,\rho}(J)$ which implies that $u^* \in PC_{\gamma,\rho}^{\gamma}(J)$. As a consequence of Steps 1 to 4 together with Theorem 4.13, we can conclude that the problem (4.35)–(4.37) has at least one solution in $PC_{\gamma,\rho}^{\gamma}(J)$. □

Our second existence result for the problem (4.35)–(4.37) is based on Darbo's fixed point theorem.

Theorem 4.14 *Assume (4.14.1)–(4.14.5) and (4.39) hold. Then the problem (4.35)–(4.37) has at least one solution in $PC_{\gamma,\rho}^{\gamma}(J)$.*

Proof Consider the operator Ψ defined in (4.38). We know that $\Psi : B_R \longrightarrow B_R$ is bounded and continuous and that $\Psi(B_R)$ is equicontinuous, we need to prove that the operator Ψ is a $\mathfrak{L}$-contraction.

Let $D \subset B_R$ and $t \in J$. Then we have

$$\mu\left(\left(\frac{t^\rho-t_k^\rho}{\rho}\right)^{1-\gamma}(\Psi D)(t)\right) = \mu\left(\left(\frac{t^\rho-t_k^\rho}{\rho}\right)^{1-\gamma}(\Psi u)(t) : u \in D\right)$$

$$\leq \frac{1}{\Gamma(\gamma)}\left[\sum_{a<t_k<t}\eta^*\mu\left(\left\{\left(\frac{t^\rho-t_k^\rho}{\rho}\right)^{1-\gamma}u(t), u \in D\right\}\right)\right.$$

$$+ \sum_{a<t_k<t}\left\{\left({}^\rho\mathcal{J}_{(t_{k-1})^+}^{1-\gamma+\alpha}p^*\mu\left(\left(\frac{s^\rho-t_k^\rho}{\rho}\right)^{1-\gamma}u(s)\right)\right)(t_k), u \in D\right\}\right]$$

$$+ \left(\frac{b^\rho-a^\rho}{\rho}\right)^{1-\gamma}\left\{\left({}^\rho\mathcal{J}_{t_k^+}^{\alpha}p^*\mu\left(\left(\frac{s^\rho-t_k^\rho}{\rho}\right)^{1-\gamma}u(s)\right)\right)(t), u \in D\right\}.$$

By Lemma 2.19, we have

$$\mu_{PC_{\gamma,\rho}}(\Psi D) \leq \left[\frac{m\eta^*}{\Gamma(\gamma)} + \left(\frac{p^*}{\Gamma(\alpha+1)} + \frac{mp^*}{\Gamma(\gamma)\Gamma(2-\gamma+\alpha)}\right)\right.$$

$$\times \left.\left(\frac{b^\rho-a^\rho}{\rho}\right)^{1-\gamma+\alpha}\right]\mu_{PC_{\gamma,\rho}}(D).$$

Therefore,

$$\mu_{PC_{\gamma,\rho}}(\Psi D) \leq \mathfrak{L}\mu_{PC_{\gamma,\rho}}(D).$$

So, by (4.39), the operator Ψ is a $\mathfrak{L}$-contraction.

As a consequence of Theorem 2.48 and using Step 4 of the last result, we deduce that Ψ has a fixed point which is a solution of the problem (4.35)–(4.37). □

4.3.2 Ulam-Type Stability

Now, we consider the Ulam stability for problem (4.35)–(4.37). Let $u \in PC_{\gamma,\rho}(J)$, $\epsilon > 0$, $\tau > 0$, and $\theta : J \longrightarrow [0, \infty)$ be a continuous function. We consider the following inequality:

$$\begin{cases} \left\|\left({}^\rho\mathcal{D}_{t_k^+}^{\alpha,\beta}u\right)(t) - f\left(t, u(t), \left({}^\rho\mathcal{D}_{t_k^+}^{\alpha,\beta}u\right)(t)\right)\right\| \leq \epsilon\theta(t), t \in J_k, k = 0, \ldots, m, \\[4mm] \left\|\left({}^\rho\mathcal{J}_{t_k^+}^{1-\gamma}u\right)(t_k^+) - \left({}^\rho\mathcal{J}_{t_{k-1}^+}^{1-\gamma}u\right)(t_k^-) - \varpi_k(u(t_k^-))\right\| \leq \epsilon\tau, k = 1, \ldots, m. \end{cases}$$

$$(4.40)$$

Definition 2 ([156]) Problem (4.35)–(4.37) is Ulam-Hyers-Rassias (U-H-R) stable with respect to (θ, τ) if there exists a real number $a_{f,m,\theta} > 0$ such that for each $\epsilon > 0$ and for each solution $u \in PC_{\gamma,\rho}(J)$ of inequality (4.40) there exists a solution $w \in PC_{\gamma,\rho}(J)$ of (4.35)–(4.37) with

$$\|u(t) - w(t)\| \leq \epsilon a_{f,m,\theta}(\theta(t) + \tau), \qquad t \in J.$$

Remark 4.15 ([156]) A function $u \in PC_{\gamma,\rho}(J)$ is a solution of inequality (4.40) if and only if there exist $\sigma \in PC_{\gamma,\rho}(J)$ and a sequence $\sigma_k, k = 0, \ldots, m$ such that

1. $\|\sigma(t)\| \leq \epsilon\theta(t)$ and $\|\sigma_k\| \leq \epsilon\tau, t \in J_k, k = 1, \ldots, m$;
2. $\left({}^{\rho}\mathcal{D}^{\alpha,\beta}_{t^+_k} u\right)(t) = f\left(t, u(t), \left({}^{\rho}\mathcal{D}^{\alpha,\beta}_{t^+_k} u\right)(t)\right) + \sigma(t), t \in J_k, k = 0, \ldots, m$;
3. $\left({}^{\rho}\mathcal{J}^{1-\gamma}_{t^+_k} u\right)(t^+_k) = \left({}^{\rho}\mathcal{J}^{1-\gamma}_{t^+_{k-1}} u\right)(t^-_k) + \varpi_k(u(t^-_k)) + \sigma_k, k = 1, \ldots, m$.

Theorem 4.16 *Assume that in addition to (4.14.1)–(4.14.5) and (4.39), the following hypothesis hold.*

(4.18.1) There exist a nondecreasing function $\theta \in PC_{\gamma,\rho}(J)$ and $\lambda_\theta > 0$ such that for each $t \in J$, we have

$$({}^{\rho}\mathcal{J}^{\alpha}_{a^+}\theta)(t) \leq \lambda_\theta\theta(t).$$

(4.18.1) There exists a continuous function $\chi : [a, b] \longrightarrow [0, \infty)$ such that for each $t \in J_k; k = 0, \ldots, m$, we have

$$p(t) \leq \chi(t)\theta(t).$$

Then Eq. (4.35) is U-H-R stable with respect to (θ, τ).

Set $\chi^* = \sup_{t \in [a,b]} \chi(t)$.

Proof Consider the operator Ψ defined in (4.38). Let $u \in PC_{\gamma,\rho}(J)$ be a solution of inequality (4.40), and let us assume that w is the unique solution of the problem

$$\begin{cases} \left({}^\rho \mathcal{D}_{t_k^+}^{\alpha,\beta} w \right)(t) = f\left(t, w(t), \left({}^\rho \mathcal{D}_{t_k^+}^{\alpha,\beta} w \right)(t) \right); \ t \in J_k, \ k = 0, \ldots, m, \\[2mm] \left({}^\rho \mathcal{J}_{t_k^+}^{1-\gamma} w \right)(t_k^+) = \left({}^\rho \mathcal{J}_{t_{k-1}^+}^{1-\gamma} w \right)(t_k^-) + \varpi_k(w(t_k^-)); \ k = 1, \ldots, m, \\[2mm] c_1 \left({}^\rho \mathcal{J}_{a^+}^{1-\gamma} w \right)(a^+) + c_2 \left({}^\rho \mathcal{J}_{t_m}^{1-\gamma} w \right)(b) = c_3, \\[2mm] \left({}^\rho \mathcal{J}_{t_k^+}^{1-\gamma} w \right)(t_k^+) = \left({}^\rho \mathcal{J}_{t_k^+}^{1-\gamma} u \right)(t_k^+); \ k = 0, \ldots, m. \end{cases}$$

By Lemma 2.38, we obtain for each $t \in (a, b]$

$$w(t) = \frac{\left({}^\rho \mathcal{J}_{t_k^+}^{1-\gamma} w \right)(t_k^+)}{\Gamma(\gamma)} \left(\frac{t^\rho - t_k^\rho}{\rho} \right)^{\gamma-1} + \left({}^\rho \mathcal{J}_{t_k^+}^{\alpha} h \right)(t) \qquad t \in J_k, k = 0, \ldots, m,$$

where $h : (a, b] \to E$ be a function satisfying the functional equation

$$h(t) = f(t, w(t), h(t)).$$

Since u is a solution of the inequality (4.40), by Remark 4.15, we have

$$\begin{cases} \left({}^\rho \mathcal{D}_{t_k^+}^{\alpha,\beta} u \right)(t) = f\left(t, u(t), \left({}^\rho \mathcal{D}_{t_k^+}^{\alpha,\beta} u \right)(t) \right) + \sigma(t), t \in J_k, k = 0, \ldots, m; \\[2mm] \left({}^\rho \mathcal{J}_{t_k^+}^{1-\gamma} u \right)(t_k^+) = \left({}^\rho \mathcal{J}_{t_{k-1}^+}^{1-\gamma} u \right)(t_k^-) + \varpi_k(u(t_k^-)) + \sigma_k, k = 1, \ldots, m. \end{cases} \qquad (4.41)$$

Clearly, the solution of (4.41) is given by

$$\begin{aligned} u(t) = {} & \frac{1}{\Gamma(\gamma)} \left(\frac{t^\rho - t_k^\rho}{\rho} \right)^{\gamma-1} \left[\left({}^\rho \mathcal{J}_{a^+}^{1-\gamma} u \right)(a^+) + \sum_{a < t_k < t} \varpi_k(u(t_k^-)) + \sum_{a < t_k < t} \sigma_k \right. \\ & \left. + \sum_{a < t_k < t} \left({}^\rho \mathcal{J}_{(t_{k-1})^+}^{1-\gamma+\alpha} g \right)(t_k) + \sum_{a < t_k < t} \left({}^\rho \mathcal{J}_{(t_{k-1})^+}^{1-\gamma+\alpha} \sigma \right)(t_k) \right] \\ & + \left({}^\rho \mathcal{J}_{t_k^+}^{\alpha} g \right)(t) + \left({}^\rho \mathcal{J}_{t_k^+}^{\alpha} \sigma \right)(t) \qquad t \in J_k, k = 0, \ldots, m, \end{aligned}$$

where $g : (a, b] \to E$ be a function satisfying the functional equation

$$g(t) = f(t, u(t), g(t)).$$

We have for each $t \in J_k, k = 0, \ldots, m,$

$$\begin{aligned} \left({}^\rho \mathcal{J}_{t_k^+}^{1-\gamma} u \right)(t_k^+) = {} & \left({}^\rho \mathcal{J}_{a^+}^{1-\gamma} u \right)(a^+) + \sum_{a < t_k < t} \varpi_k(u(t_k^-)) + \sum_{a < t_k < t} \sigma_k \\ & + \sum_{a < t_k < t} \left({}^\rho \mathcal{J}_{(t_{k-1})^+}^{1-\gamma+\alpha} g \right)(t_k) + \sum_{a < t_k < t} \left({}^\rho \mathcal{J}_{(t_{k-1})^+}^{1-\gamma+\alpha} \sigma \right)(t_k). \end{aligned}$$

Hence, for each $t \in J$, we have

$$\|u(t) - w(t)\| \leq \left({}^{\rho}\mathcal{J}_{t_k^+}^{\alpha}|g(s) - h(s)|\right)(t) + \left({}^{\rho}\mathcal{J}_{t_k^+}^{\alpha}|\sigma(s)|\right)(t).$$

Thus,

$$
\begin{aligned}
\|u(t) - w(t)\| &\leq \left({}^{\rho}\mathcal{J}_{a^+}^{\alpha}\|g(s) - h(s)\|\right)(t) + \left({}^{\rho}\mathcal{J}_{a^+}^{\alpha}\|\sigma(s)\|\right) \\
&\leq \epsilon\lambda_\theta\theta(t) + \int_a^t s^{\rho-1}\left(\frac{t^\rho - s^\rho}{\rho}\right)^{\alpha-1}\frac{2\chi(t)\theta(t)}{\Gamma(\gamma)}ds \\
&\leq \epsilon\lambda_\theta\theta(t) + 2\chi^*\left({}^{\rho}\mathcal{J}_{a^+}^{\alpha}\theta\right)(t) \\
&\leq (\epsilon + 2\chi^*)\lambda_\theta\theta(t) \\
&\leq (1 + \frac{2\chi^*}{\epsilon})\lambda_\theta\epsilon(\tau + \theta(t)) \\
&\leq a_\theta\epsilon(\tau + \theta(t)),
\end{aligned}
$$

where $a_\theta = (1 + \frac{2\chi^*}{\epsilon})\lambda_\theta$. Hence, Eq. (4.35) is U-H-R stable with respect to (θ, τ). $\qquad\square$

4.3.3 Examples

Let

$$E = l^1 = \left\{u = (u_1, u_2, \cdots, u_n, \cdots), \sum_{n=1}^{\infty}|u_n| < \infty\right\}$$

be the Banach space with the norm

$$\|u\| = \sum_{n=1}^{\infty}|u_n|.$$

Example 4.17 Consider the following impulsive boundary value problem of generalized Hilfer fractional differential equation

$$\left({}^{1}\mathcal{D}_{t_k^+}^{\frac{1}{2},0}u_n\right)(t) = \frac{3t^2 - 20}{213e^{-t+3}(1 + |u_n(t)| + |{}^{1}\mathcal{D}_{t_k^+}^{\frac{1}{2},0}u_n(t)|)}, \quad t \in J_k, k = 0, \cdots, 9,$$

(4.42)

$$\left({}^{1}\mathcal{J}_{t_k^+}^{\frac{1}{2}}u_n\right)(t_k^+) - \left({}^{1}\mathcal{J}_{t_{(k-1)^+}}^{\frac{1}{2}}u_n\right)(t_k^-) = \frac{|u_n(t_k^-)|}{10(k+3) + |u_n(t_k^-)|}, \quad k = 1, \cdots, 9, \quad (4.43)$$

$$\left({}^{1}\mathcal{J}_{1^+}^{\frac{1}{2}}u_n\right)(1^+) + 2\left({}^{1}\mathcal{J}_{\frac{9}{5}+}^{\frac{1}{2}}u_n\right)(3) = 0, \quad (4.44)$$

where $J_k = (t_k, t_{k+1}], t_k = 1 + \dfrac{k}{5}$ for $k = 0, \cdots, 9, m = 9, a = t_0 = 1$, and $b = t_{10} = 3$.

Set

$$f(t, u, w) = \frac{3t^2 - 20}{213e^{-t+3}(1 + \|u\| + \|w\|)}, \quad t \in (1, 3], \ u, w \in E.$$

We have

$$PC_{\gamma,\rho}^{\beta(1-\alpha)}([1, 3]) = PC_{\frac{1}{2},1}^{0}([1, 3]) = \left\{g : (1, 3] \to \mathbb{R} : (\sqrt{t - t_k})g \in PC([1, 3])\right\},$$

with $\gamma = \alpha = \frac{1}{2}, \rho = 1, \beta = 0$, and $k = 0, \cdots, 9$. Clearly, the continuous function $f \in PC_{\frac{1}{2},1}^{0}([1, 2])$.

Hence, the condition (4.14.1) is satisfied.

For each $u, w \in E$ and $t \in (1, 3]$:

$$\|f(t, u, w)\| \le \frac{3t^2 - 20}{213e^{-t+3}}.$$

Hence, condition (4.14.2) is satisfied with $p^* = \dfrac{7}{213}$.

And let

$$\varpi_k(u) = \frac{\|u\|}{10(k + 3) + \|u\|}, k = 1, \cdots, 9, u \in E.$$

Let $u \in E$. Then we have

$$\|\varpi_k(u)\| \le \frac{1}{40}\|u\|, k = 1, \cdots, 9,$$

and so the condition (4.14.4) is satisfied with $\eta^* = \dfrac{1}{40}$.

The condition (4.39) of Theorem 4.13 is satisfied for

$$\begin{aligned}
\mathcal{L} &:= \frac{m\eta^*}{\Gamma(\gamma)} + \left(\frac{p^*}{\Gamma(\alpha + 1)} + \frac{mp^*}{\Gamma(\gamma)\Gamma(2 - \gamma + \alpha)}\right)\left(\frac{b^\rho - a^\rho}{\rho}\right)^{1-\gamma+\alpha} \\
&= \frac{9}{40\sqrt{\pi}} + 2\left(\frac{17}{213\sqrt{\pi}} + \frac{63}{213\Gamma(2)\sqrt{\pi}}\right) \\
&\approx 0.55074703829 < 1.
\end{aligned}$$

Then the problem (4.42)–(4.44) has at least one solution in $PC_{\frac{1}{2},1}^{\frac{1}{2}}([1, 3])$.

Example 4.18 Let the following impulsive anti-periodic boundary value problem

$$\left(\tfrac{1}{2}D_{t_k^+}^{\frac{1}{2},0}u_n\right)(t) = \frac{(3t^3 + 5e^{-3})|u_n(t)|}{144e^{-t+e}(1 + \|u(t)\| + \|\tfrac{1}{2}D_{t_k^+}^{\frac{1}{2},0}u(t)\|)}, \quad \text{for each } t \in J_0 \cup J_1, \quad (4.45)$$

$$\left(\tfrac{1}{2}\mathcal{J}_{2^+}^{\frac{1}{2}}u_n\right)(2^+) - \left(\tfrac{1}{2}\mathcal{J}_{1^+}^{\frac{1}{2}}u_n\right)(2^-) = \frac{|u_n(2^-)|}{77e^{-t+4}+2}, \tag{4.46}$$

$$\left(\tfrac{1}{2}\mathcal{J}_{1^+}^{\frac{1}{2}}u\right)(1^+) = -\left(\tfrac{1}{2}\mathcal{J}_{2^+}^{\frac{1}{2}}u\right)(e), \tag{4.47}$$

where $J_0 = (1, 2]$, $J_1 = (2, e]$, $t_1 = 2$, $m = 1$, $a = t_0 = 1$, and $b = t_2 = e$.
Set

$$f(t, u, w) = \frac{(3t^3 + 5e^{-3})\|u\|}{144e^{-t+e}(1 + \|u\| + \|w\|)}, \ t \in (1, e], \ u, w \in E.$$

We have

$$PC_{\gamma,\rho}^{\beta(1-\alpha)}([1, 2]) = PC_{\frac{1}{2},\frac{1}{2}}^0([1, e])$$

$$= \left\{ g : (1, e] \to E : \sqrt{2}(\sqrt{t} - \sqrt{t_k})^{\frac{1}{2}} g \in C([1, e]) \right\},$$

with $\gamma = \alpha = \frac{1}{2}$, $\rho = \frac{1}{2}$, $\beta = 0$, and $k \in \{0, 1\}$. Clearly, the continuous function $f \in PC_{\frac{1}{2},\frac{1}{2}}^0([1, e])$.
Hence, the condition (4.14.1) is satisfied.
For each $u, w \in E$ and $t \in (1, e]$:

$$\|f(t, u, w)\| \le \frac{(3t^3 + 5e^{-3})}{144e^{-t+e}}.$$

Hence, condition (4.14.2) is satisfied with

$$p(t) = \frac{(3t^3 + 5e^{-3})}{144e^{-t+e}},$$

and

$$p^* = \frac{(3e^3 + 5e^{-3})}{144}.$$

And let

$$\varpi_1(u) = \frac{\|u\|}{77e^{-t+4}+2}, \ u \in E.$$

Let $u \in E$. Then we have

$$\|\varpi_k(u)\| \le \frac{1}{77e^{-t+4}+2}\|u\|,$$

and so the condition (4.14.4) is satisfied with $\eta^* = \frac{1}{77e^{4-e}+2}$.

The condition (4.39) of Theorem 4.13 is satisfied for

$$
\mathfrak{L} := \frac{m\eta^*}{\Gamma(\gamma)} + \left(\frac{p^*}{\Gamma(\alpha+1)} + \frac{mp^*}{\Gamma(\gamma)\Gamma(2-\gamma+\alpha)}\right)\left(\frac{b^\rho - a^\rho}{\rho}\right)^{1-\gamma+\alpha}
$$
$$
= \frac{1}{(77e^{4-e}+2)\sqrt{\pi}} + (2\sqrt{e}-2)\left(\frac{6e^3+10e^{-3}}{144\sqrt{\pi}} + \frac{3e^3+5e^{-3}}{144\sqrt{\pi}\Gamma(2)}\right)
$$
$$
\approx 0.92473323802 < 1.
$$

Then the problem (4.45)–(4.47) has at least one solution in $PC_{\frac{1}{2},\frac{1}{2}}^{\frac{1}{2}}([1,e])$. Also, hypothesis (4.18.1) is satisfied with $\tau = 1$, $\theta(t) = e^3$, and $\lambda_\theta = 3$. Indeed, for each $t \in (1,e]$, we get

$$
(\tfrac{1}{2}\mathcal{J}_{1+}^{\frac{1}{2}}\theta)(t) \le \frac{2e^3}{\Gamma(\frac{3}{2})} \le \lambda_\theta\theta(t).
$$

Let the function $\chi : [1,e] \longrightarrow [0,\infty)$ be defined by

$$
\chi(t) = \frac{(3e^{-3}t^3 + 5e^{-6})}{144e^{-t+e}},
$$

then, for each $t \in (1,e]$, we have

$$
p(t) = \chi(t)\theta(t),
$$

with $\chi^* = p^*e^{-3}$. Hence, the condition (4.18.2) is satisfied. Consequently, Theorem 4.16 implies that Eq. (4.45) is U-H-R stable.

4.4   Notes and Remarks

The results of this chapter are taken from the papers of Salim et al. [125, 133]. The monographs [7, 8, 14, 27, 43, 81, 98, 115, 151, 159, 160], and the papers [17, 19, 34, 48, 49, 83, 84, 88, 132] provide more important conclusions and analyses about the subject.

Fractional Differential Equations with Non-Instantaneous Impulses

5

5.1 Introduction and Motivations

The present chapter deals with some existence, uniqueness, and Ulam stability results for a class of initial and boundary value problems for nonlinear implicit fractional differential equations with non-instantaneous impulses and generalized Hilfer-type fractional derivative. The tools employed are some suitable fixed point theorems combined with the technique of measure of noncompactness. We provide illustrations to demonstrate the applicability of our results for each section.

The outcome of our study in this chapter can be considered as a partial continuation of the problems raised recently in the following:

- The monographs of Abbas et al. [7, 8, 14], Ahmad et al. [25], and Baleanu et al. [43], and the papers of Ahmed et al. [28], which deal with various linear and nonlinear initial and boundary value problems for fractional differential equations involving different kinds of fractional derivatives.
- The monographs of Abbas et al. [7], Agarwal et al. [24], Benchohra et al. [50], and Stamova et al. [150], and the papers of Abbas et al. [1–6, 8, 15], Bai et al. [39], Hernández et al. [78], Kong et al. [87], and Wang et al. [155, 157], where the authors investigated the class of problems for fractional differential equations with impulsive conditions, and the books [69, 154], where different topics on the qualitative properties of solutions are considered.
- The monographs of Abbas et al. [7, 13], and the papers of Abbas et al. [10, 12] and Benchohra et al. [51, 52]; in it, considerable attention has been given to the study of the Ulam-Hyers and Ulam-Hyers-Rassias stability of various classes of functional equations.

© The Author(s), under exclusive license to Springer Nature Switzerland AG 2023   117
M. Benchohra et al., *Advanced Topics in Fractional Differential Equations*,
Synthesis Lectures on Mathematics & Statistics,
https://doi.org/10.1007/978-3-031-26928-8_5

5.2   Initial Value Problem for Nonlinear Implicit Generalized Hilfer-Type Fractional Differential Equations

In this section, we establish existence results to the initial value problem with nonlinear implicit generalized Hilfer-type fractional differential equation with non-instantaneous impulses:

$$\left({}^\rho D_{s_k^+}^{\alpha,\beta} u \right)(t) = f\left(t, u(t), \left({}^\rho D_{s_k^+}^{\alpha,\beta} u \right)(t) \right); \ t \in I_k, \ k = 0, \ldots, m, \tag{5.1}$$

$$u(t) = g_k(t, u(t)); \ t \in \tilde{I}_k, \ k = 1, \ldots, m, \tag{5.2}$$

$$\left({}^\rho J_{a^+}^{1-\gamma} u \right)(a^+) = \phi_0, \tag{5.3}$$

where ${}^\rho D_{s_k^+}^{\alpha,\beta}$, ${}^\rho J_{a^+}^{1-\gamma}$ are the generalized Hilfer fractional derivative of order $\alpha \in (0,1)$ and type $\beta \in [0,1]$ and generalized fractional integral of order $1 - \gamma$, $(\gamma = \alpha + \beta - \alpha\beta)$, respectively, $\phi_0 \in \mathbb{R}$, $I_k := (s_k, t_{k+1}]$; $k = 0, \ldots, m$, $\tilde{I}_k := (t_k, s_k]$; $k = 1, \ldots, m$, $a = t_0 = s_0 < t_1 \le s_1 < t_2 \le s_2 < \cdots \le s_{m-1} < t_m \le s_m < t_{m+1} = b < \infty$, $u(t_k^+) = \lim_{\epsilon \to 0^+} u(t_k + \epsilon)$ and $u(t_k^-) = \lim_{\epsilon \to 0^-} u(t_k + \epsilon)$ represent the right- and left-hand limits of $u(t)$ at $t = t_k$, $f : J \times \mathbb{R} \times \mathbb{R} \to \mathbb{R}$ is a given function, and $g_k : \tilde{I}_k \times \mathbb{R} \to \mathbb{R}$; $k = 1, \ldots, m$ are given continuous functions such that $\left({}^\rho J_{s_k^+}^{1-\gamma} g_k \right)(t, u(t)) \big|_{t=s_k} = \phi_k \in \mathbb{R}$.

5.2.1   Existence Results

Consider the Banach space

$$PC_{\gamma,\rho}(J) = \Big\{ u : J \to \mathbb{R} : u \in C_{\gamma,\rho}(I_k, \mathbb{R}); k = 0, \ldots, m, \text{ and }$$

$$u \in C(\tilde{I}_k, \mathbb{R}); k = 1, \ldots, m, \text{ and there exist } u(t_k^-), u(t_k^+),$$

$$u(s_k^-), \text{ and } u(s_k^+) \text{ with } u(t_k^-) = u(t_k) \Big\},$$

and

$$PC_{\gamma,\rho}^n(J) = \Big\{ u \in PC^{n-1}(J) : u^{(n)} \in PC_{\gamma,\rho}(J) \Big\}, n \in \mathbb{N},$$

$$PC_{\gamma,\rho}^0(J) = PC_{\gamma,\rho}(J),$$

with the norm

$$\|u\|_{PC_{\gamma,\rho}}$$

$$= \max \left\{ \max_{k=0,\dots,m} \left\{ \sup_{t\in[s_k,t_{k+1}]} \left| \left(\frac{t^\rho - s_k^\rho}{\rho} \right)^{1-\gamma} u(t) \right| \right\}, \max_{k=1,\dots,m} \left\{ \sup_{t\in[t_k,s_k]} |u(t)| \right\} \right\}.$$

We define the space

$$PC_{\gamma,\rho}^\gamma(J) = \Big\{ u : J \to \mathbb{R} : u \in C_{\gamma,\rho}^\gamma(I_k, \mathbb{R}); \, k = 0, \dots, m, \text{ and}$$

$$u \in C(\tilde{I}_k, \mathbb{R}); \, k = 1, \dots, m, \text{ and there exist } u(t_k^-), u(t_k^+),$$

$$u(s_k^-), \text{ and } u(s_k^+) \text{ with } u(t_k^-) = u(t_k) \Big\}.$$

We consider the following linear fractional differential equation:

$$\left({}^\rho\mathcal{D}_{s_k^+}^{\alpha,\beta} u \right)(t) = \psi(t), \; t \in I_k, \; k = 0, \dots, m, \tag{5.4}$$

where $0 < \alpha < 1, 0 \le \beta \le 1, \rho > 0$, with the conditions

$$u(t) = g_k(t, u(t)); \; t \in \tilde{I}_k, \; k = 1, \dots, m, \tag{5.5}$$

and

$$\left({}^\rho\mathcal{J}_{a^+}^{1-\gamma} u \right)(a^+) = \phi_0, \tag{5.6}$$

where $\gamma = \alpha + \beta - \alpha\beta$, $\phi_0 \in \mathbb{R}$, and $\phi^* = max\{|\phi_k| : k = 0, \dots, m\}$. The following the-
orem shows that the problem (5.4)–(5.6) has a unique solution given by

$$u(t) = \begin{cases} \dfrac{\phi_k}{\Gamma(\gamma)} \left(\dfrac{t^\rho - s_k^\rho}{\rho} \right)^{\gamma-1} + \left({}^\rho\mathcal{J}_{s_k^+}^{\alpha} \psi \right)(t) & \text{if } t \in I_k, \; k = 0, \dots, m, \\[4mm] u(t) = g_k(t, u(t)) & \text{if } t \in \tilde{I}_k, \; k = 1, \dots, m. \end{cases} \tag{5.7}$$

Theorem 5.1 *Let* $\gamma = \alpha + \beta - \alpha\beta$*, where* $0 < \alpha < 1$ *and* $0 \le \beta \le 1$*. If* $\psi : I_k \to \mathbb{R}; k = 0, \dots, m$*, is a function such that* $\psi(\cdot) \in C_{\gamma,\rho}(I_k)$*, then* $u \in PC_{\gamma,\rho}^\gamma(J)$ *satisfies the problem* (5.4)–(5.6) *if and only if it satisfies* (5.7).

Proof Assume u satisfies (5.4)–(5.6). If $t \in I_0$, then

$$\left({}^\rho\mathcal{D}_{a^+}^{\alpha,\beta} u \right)(t) = \psi(t).$$

Lemma 2.38 implies we have a solution that can be written as

$$u(t) = \frac{\left({}^{\rho}\mathcal{J}_{a^+}^{1-\gamma}u\right)(a)}{\Gamma(\gamma)} \left(\frac{t^\rho - a^\rho}{\rho}\right)^{\gamma-1} + \frac{1}{\Gamma(\alpha)} \int_a^t \left(\frac{t^\rho - s^\rho}{\rho}\right)^{\alpha-1} s^{\rho-1}\psi(s)ds.$$

If $t \in \tilde{I}_1$, then we have $u(t) = g_1(t, u(t))$.
If $t \in I_1$, then Lemma 2.38 implies

$$u(t) = \frac{\left({}^{\rho}\mathcal{J}_{s_1^+}^{1-\gamma}u\right)(s_1)}{\Gamma(\gamma)} \left(\frac{t^\rho - s_1^\rho}{\rho}\right)^{\gamma-1} + \frac{1}{\Gamma(\alpha)} \int_{s_1}^t \left(\frac{t^\rho - s^\rho}{\rho}\right)^{\alpha-1} s^{\rho-1}\psi(s)ds$$

$$= \frac{\phi_1}{\Gamma(\gamma)} \left(\frac{t^\rho - s_1^\rho}{\rho}\right)^{\gamma-1} + \left({}^{\rho}\mathcal{J}_{s_1^+}^{\alpha}\psi\right)(t).$$

If $t \in \tilde{I}_2$, then we have $u(t) = g_2(t, u(t))$.
If $t \in I_2$, then Lemma 2.38 implies

$$u(t) = \frac{\left({}^{\rho}\mathcal{J}_{s_2^+}^{1-\gamma}u\right)(s_2)}{\Gamma(\gamma)} \left(\frac{t^\rho - s_2^\rho}{\rho}\right)^{\gamma-1} + \frac{1}{\Gamma(\alpha)} \int_{s_2}^t \left(\frac{t^\rho - s^\rho}{\rho}\right)^{\alpha-1} s^{\rho-1}\psi(s)ds$$

$$= \frac{\phi_2}{\Gamma(\gamma)} \left(\frac{t^\rho - s_2^\rho}{\rho}\right)^{\gamma-1} + \left({}^{\rho}\mathcal{J}_{s_2^+}^{\alpha}\psi\right)(t).$$

Repeating the process in this way, the solution $u(t)$ for $t \in J$ can be written as

$$u(t) = \begin{cases} \frac{\phi_k}{\Gamma(\gamma)} \left(\frac{t^\rho - s_k^\rho}{\rho}\right)^{\gamma-1} + \left({}^{\rho}\mathcal{J}_{s_k^+}^{\alpha}\psi\right)(t) & \text{if } t \in I_k, k = 0, \ldots, m, \\[2ex] u(t) = g_k(t, u(t)) & \text{if } t \in \tilde{I}_k, \ k = 1, \ldots, m. \end{cases}$$

Reciprocally, for $t \in I_0$, applying ${}^{\rho}\mathcal{J}_{a^+}^{1-\gamma}$ on both sides of (5.7) and using Lemma 2.19 and Theorem 2.14, we get

$$\left({}^{\rho}\mathcal{J}_{a^+}^{1-\gamma}u\right)(t) = \phi_0 + \left({}^{\rho}\mathcal{J}_{a^+}^{1-\gamma+\alpha}\psi\right)(t). \tag{5.8}$$

Next, taking the limit $t \to a^+$ of (5.8) and using Lemma 2.24, with $1 - \gamma < 1 - \gamma + \alpha$, we obtain

$$\left({}^{\rho}\mathcal{J}_{a^+}^{1-\gamma}u\right)(a^+) = \phi_0, \tag{5.9}$$

which shows that the initial condition $\left({}^{\rho}\mathcal{J}_{a^+}^{1-\gamma}u\right)(a^+) = \phi_0$ is satisfied. Next, for $t \in I_k; k = 0, \ldots, m$, apply operator ${}^{\rho}\mathcal{D}_{s_k^+}^{\gamma}$ on both sides of (5.7). Then, from Lemmas 2.19 and 2.33, we obtain

$$({}^\rho \mathcal{D}^\gamma_{s_k^+} u)(t) = \left({}^\rho \mathcal{D}^{\beta(1-\alpha)}_{s_k^+} \psi\right)(t). \tag{5.10}$$

Since $u \in C^\gamma_{\gamma,\rho}(I_k)$ and by definition of $C^\gamma_{\gamma,\rho}(I_k)$, we have ${}^\rho \mathcal{D}^\gamma_{s_k^+} u \in C_{\gamma,\rho}(I_k)$, then (5.10) implies that

$$({}^\rho \mathcal{D}^\gamma_{s_k^+} u)(t) = \left(\delta_\rho \, {}^\rho \mathcal{J}^{1-\beta(1-\alpha)}_{s_k^+} \psi\right)(t) = \left({}^\rho \mathcal{D}^{\beta(1-\alpha)}_{s_k^+} \psi\right)(t) \in C_{\gamma,\rho}(I_k). \tag{5.11}$$

As $\psi(\cdot) \in C_{\gamma,\rho}(I_k)$ and from Lemma 2.23, follows

$$\left({}^\rho \mathcal{J}^{1-\beta(1-\alpha)}_{s_k^+} \psi\right) \in C_{\gamma,\rho}(I_k), k = 0, \ldots, m. \tag{5.12}$$

From (5.11), (5.12), and by the definition of the space $C^n_{\gamma,\rho}(I_k)$, we obtain

$$\left({}^\rho \mathcal{J}^{1-\beta(1-\alpha)}_{s_k^+} \psi\right) \in C^1_{\gamma,\rho}(I_k), k = 0, \ldots, m.$$

Applying operator ${}^\rho \mathcal{J}^{\beta(1-\alpha)}_{s_k^+}$ on both sides of (5.10) and using Lemmas 2.32 and 2.24 and Property 2.22, we have

$$\left({}^\rho \mathcal{D}^{\alpha,\beta}_{s_k^+} u\right)(t) = {}^\rho \mathcal{J}^{\beta(1-\alpha)}_{s_k^+} \left({}^\rho \mathcal{D}^\gamma_{s_k^+} u\right)(t)$$

$$= \psi(t) - \frac{\left({}^\rho \mathcal{J}^{1-\beta(1-\alpha)}_{s_k^+} \psi\right)(s_k)}{\Gamma(\beta(1-\alpha))} \left(\frac{t^\rho - s_k^\rho}{\rho}\right)^{\beta(1-\alpha)-1}$$

$$= \psi(t),$$

that is, (5.4) holds.
Also, we can easily show that

$$u(t) = g_k(t, u(t_k^-)); \ t \in \tilde{I}_k, \ k = 1, \ldots, m.$$

This completes the proof.   □

As a consequence of Theorem 5.1, we have the following result.

Lemma 5.2 *Let* $\gamma = \alpha + \beta - \alpha\beta$ *where* $0 < \alpha < 1$ *and* $0 \le \beta \le 1$, *and* $k = 0, \ldots, m$, *let* $f : J \times \mathbb{R} \times \mathbb{R} \to \mathbb{R}$ *be a function such that* $f(\cdot, u(\cdot), w(\cdot)) \in C_{\gamma,\rho}(I_k)$, *for any* $u, w \in PC_{\gamma,\rho}(J)$. *If* $u \in PC^\gamma_{\gamma,\rho}(J)$, *then* u *satisfies the problem* (5.1)–(5.3) *if and only if* u *is the fixed point of the operator* $\Psi : PC_{\gamma,\rho}(J) \to PC_{\gamma,\rho}(J)$ *defined by*

$$\Psi u(t) = \begin{cases} \dfrac{\phi_k}{\Gamma(\gamma)} \left(\dfrac{t^\rho - s_k^\rho}{\rho} \right)^{\gamma-1} + \left({}^\rho \mathcal{J}^\alpha_{s_k^+} h \right)(t) & if \ t \in I_k, \ k = 0, \ldots, m, \\[4mm] g_k(t, u(t)) & if \ t \in \tilde{I}_k, \ k = 1, \ldots, m. \end{cases} \tag{5.13}$$

where $h \in C_{\gamma,\rho}(I_k)$, $k = 0, \ldots, m$ be a function satisfying the functional equation

$$h(t) = f(t, u(t), h(t)).$$

Also, by Lemma 2.23, $\Psi u \in PC_{\gamma,\rho}(J)$.

We are now in a position to state and prove our existence result for the problem (5.1)–(5.3) based on Banach's fixed point theorem. Set $\Upsilon = \frac{K}{1-M}$.

Theorem 5.3 *Suppose that the following assumptions hold.*

(5.3.1) *The function $f : I_k \times \mathbb{R} \times \mathbb{R} \to \mathbb{R}$ is continuous on I_k; $k = 0, \ldots, m$, and*

$$f(\cdot, u(\cdot), w(\cdot)) \in C^{\beta(1-\alpha)}_{\gamma,\rho}(I_k), k = 0, \ldots, m, \ \text{for any } u, w \in PC_{\gamma,\rho}(J).$$

(5.3.2) *There exist constants $\mathfrak{M}_1 > 0$ and $0 < \mathfrak{M}_2 < 1$ such that*

$$|f(t, u, w) - f(t, \bar{u}, \bar{w})| \le \mathfrak{M}_1 |u - \bar{u}| + \mathfrak{M}_2 |w - \bar{w}|$$

for any $u, w, \bar{u}, \bar{w} \in \mathbb{R}$ and $t \in I_k$, $k = 0, \ldots, m$.

(5.3.3) *The functions g_k are continuous and there exists a constant $l^* > 0$ such that $|g_k(u) - g_k(\bar{u})| \le l^* |u - \bar{u}|$ for any $u, \bar{u} \in \mathbb{R}$ and $k = 1, \ldots, m$.*

If

$$L := l^* + \frac{\Upsilon \Gamma(\gamma)}{\Gamma(\gamma + \alpha)} \left(\frac{b^\rho - a^\rho}{\rho} \right)^\alpha < 1, \tag{5.14}$$

then the problem (5.1)–(5.3) has a unique solution in $PC_{\gamma,\rho}(J)$.

Proof The proof will be given in two steps.

Step 1: We show that the operator Ψ defined in (5.13) has a unique fixed point u^* in $PC_{\gamma,\rho}(J)$. Let $u, w \in PC_{\gamma,\rho}(J)$ and $t \in J$.

For $t \in I_k, k = 0, \ldots, m$, we have

$$|\Psi u(t) - \Psi w(t)| \le \left({}^\rho \mathcal{J}^\alpha_{s_k^+} |h(s) - g(s)| \right)(t),$$

where $h, g \in C_{\gamma,\rho}(I_k)$; $k = 0, \ldots, m$, such that

$$h(t) = f(t, u(t), h(t)),$$
$$g(t) = f(t, w(t), g(t)).$$

By (5.3.2), we have

$$|h(t) - g(t)| = |f(t, u(t), h(t)) - f(t, w(t), g(t))|$$
$$\leq \mathfrak{M}_1 |u(t) - w(t)| + \mathfrak{M}_2 |h(t) - g(t)|.$$

Then,

$$|h(t) - g(t)| \leq \Upsilon |u(t) - w(t)|.$$

Therefore, for each $t \in I_k$, $k = 0, \ldots, m$,

$$|\Psi u(t) - \Psi w(t)| \leq \Upsilon \left({}^{\rho}\mathcal{J}_{s_k^+}^{\alpha} |u(s) - w(s)| \right)(t).$$

Thus

$$|\Psi u(t) - \Psi w(t)| \leq \left[\Upsilon \left({}^{\rho}\mathcal{J}_{s_k^+}^{\alpha} \left(\frac{s^{\rho} - s_k^{\rho}}{\rho} \right)^{\gamma-1} \right)(t) \right] \|u - w\|_{PC_{\gamma,\rho}}.$$

By Lemma 2.19, we have

$$|\Psi u(t) - \Psi w(t)| \leq \left[\frac{\Upsilon \Gamma(\gamma)}{\Gamma(\gamma + \alpha)} \left(\frac{t^{\rho} - s_k^{\rho}}{\rho} \right)^{\alpha+\gamma-1} \right] \|u - w\|_{PC_{\gamma,\rho}}.$$

Hence

$$\left| (\Psi u(t) - \Psi w(t)) \left(\frac{t^{\rho} - s_k^{\rho}}{\rho} \right)^{1-\gamma} \right| \leq \left[\frac{\Upsilon \Gamma(\gamma)}{\Gamma(\gamma + \alpha)} \left(\frac{t^{\rho} - s_k^{\rho}}{\rho} \right)^{\alpha} \right] \|u - w\|_{PC_{\gamma,\rho}}$$

$$\leq \left[l^* + \frac{\Upsilon \Gamma(\gamma)}{\Gamma(\gamma + \alpha)} \left(\frac{b^{\rho} - a^{\rho}}{\rho} \right)^{\alpha} \right] \|u - w\|_{PC_{\gamma,\rho}}.$$

For $t \in \tilde{I}_k$, $k = 1, \ldots, m$, we have

$$|\Psi u(t) - \Psi w(t)| \leq |(g_k(t, u(t)) - g_k(t, w(t)))|$$
$$\leq l^* \|u - w\|_{PC_{\gamma,\rho}}$$
$$\leq \left[l^* + \frac{\Upsilon \Gamma(\gamma)}{\Gamma(\gamma + \alpha)} \left(\frac{b^{\rho} - a^{\rho}}{\rho} \right)^{\alpha} \right] \|u - w\|_{PC_{\gamma,\rho}}.$$

Then, for each $t \in J$, we have

$$\|\Psi u - \Psi w\|_{PC_{\gamma,\rho}} \leq \left[l^* + \frac{\Upsilon \Gamma(\gamma)}{\Gamma(\gamma + \alpha)} \left(\frac{b^{\rho} - a^{\rho}}{\rho} \right)^{\alpha} \right] \|u - w\|_{PC_{\gamma,\rho}}.$$

By (5.14), the operator Ψ is a contraction. Hence, by Theorem 2.45, Ψ has a unique fixed point $u^* \in PC_{\gamma,\rho}(J)$.

Step 2: We show that such a fixed point $u^* \in PC_{\gamma,\rho}(J)$ is actually in $PC_{\gamma,\rho}^{\gamma}(J)$. Since u^* is the unique fixed point of operator Ψ in $PC_{\gamma,\rho}(J)$, then for each $t \in J$, we have

$$
\Psi u^*(t) = \begin{cases} \dfrac{\phi_k}{\Gamma(\gamma)} \left(\dfrac{t^\rho - s_k^\rho}{\rho} \right)^{\gamma-1} + \left({}^\rho \mathcal{J}_{s_k^+}^{\alpha} h \right)(t) & \text{if } t \in I_k, k = 0, \ldots, m, \\ g_k(t, u^*(t)) & \text{if } t \in \tilde{I}_k, k = 1, \ldots, m, \end{cases}
$$

where $h \in C_{\gamma,\rho}(I_k); k = 0, \ldots, m$, such that

$$
h(t) = f(t, u^*(t), h(t)).
$$

Applying ${}^\rho \mathcal{D}_{s_k^+}^{\gamma}$ to both sides and by Lemmas 2.19 and 2.33, we have

$$
{}^\rho \mathcal{D}_{s_k^+}^{\gamma} u^*(t) = \left({}^\rho \mathcal{D}_{s_k^+}^{\gamma} \, {}^\rho \mathcal{J}_{s_k^+}^{\alpha} f(s, u^*(s), h(s)) \right)(t)
$$
$$
= \left({}^\rho \mathcal{D}_{s_k^+}^{\beta(1-\alpha)} f(s, u^*(s), h(s)) \right)(t).
$$

Since $\gamma \geq \alpha$, by (5.3.1), the right-hand side is in $C_{\gamma,\rho}(I_k)$ and thus ${}^\rho \mathcal{D}_{s_k^+}^{\gamma} u^* \in C_{\gamma,\rho}(I_k)$. And since $g_k \in C(\tilde{I}_k, \mathbb{R}); k = 1, \ldots, m$, then $u^* \in PC_{\gamma,\rho}^{\gamma}(J)$. As a consequence of Steps 1 and 2 together with Theorem 5.3, we can conclude that the problem (5.1)–(5.3) has a unique solution in $PC_{\gamma,\rho}(J)$. □

Our second result is based on Schaefer's fixed point theorem. Set

$$
p_1^* = \sup_{t \in [a,b]} p_1(t), \quad p_2^* = \sup_{t \in [a,b]} p_2(t), \quad p_3^* = \sup_{t \in [a,b]} p_3(t) < 1.
$$

Theorem 5.4 *Assume that in addition to the hypothesis (5.3.1), the following assumptions are met.*

(5.4.1) *There exist functions $p_1, p_2, p_3 \in C([a, b], \mathbb{R}_+)$ such that*

$$
|f(t, u, w)| \leq p_1(t) + p_2(t)|u| + p_3(t)|w|,
$$

for $t \in I_k; k = 0, \ldots, m$, and $u, w \in \mathbb{R}$.

(5.4.2) *The functions g_k are continuous and there exist constants $\Phi_1, \Phi_2 > 0$ such that*

$$
|g_k(t, u)| \leq \Phi_1 |u| + \Phi_2 \text{ for each } u \in \mathbb{R}, t \in J_k', \quad k = 1, \ldots, m.
$$

If

$$max \left\{ \Phi_1, \left(\frac{p_2^* \Gamma(\gamma)}{(1 - p_3^*) \Gamma(\gamma + \alpha)} \right) \left(\frac{b^\rho - a^\rho}{\rho} \right)^\alpha \right\} < 1, \qquad (5.15)$$

then the problem (5.1)–(5.3) has at least one solution in $PC_{\gamma,\rho}(J)$.

Proof We shall use Schaefer's fixed point theorem to prove in several steps that the operator Ψ defined in (5.13) has a fixed point.

Step 1: Ψ is continuous. Let $\{u_n\}$ be a sequence such that $u_n \to u$ in $PC_{\gamma,\rho}(J)$. Then for each $t \in I_k, k = 0, \ldots, m$, we have

$$\left| ((\Psi u_n)(t) - (\Psi u)(t)) \left(\frac{t^\rho - s_k^\rho}{\rho} \right)^{1-\gamma} \right| \leq \left(\frac{t^\rho - s_k^\rho}{\rho} \right)^{1-\gamma} \left({}^\rho \mathcal{J}_{s_k^+}^\alpha |h_n(s) - h(s)| \right)(t),$$

where $h_n, h \in C_{\gamma,\rho}(I_k)$, such that

$$h_n(t) = f(t, u_n(t), h_n(t)),$$
$$h(t) = f(t, u(t), h(t)).$$

For each $t \in \tilde{I}_k, k = 1, \ldots, m$, we have

$$|\Psi u_n(t) - \Psi u(t)| \leq |(g_k(t, u_n(t)) - g_k(t, u(t)))|.$$

Since $u_n \to u$, then we get $h_n(t) \to h(t)$ as $n \to \infty$ for each $t \in J$, and since f and g_k are continuous, then we have

$$\|\Psi u_n - \Psi u\|_{PC_{\gamma,\rho}} \to 0 \text{ as } n \to \infty.$$

Step 2: We show that Ψ is the mapping of two bounded sets in $PC_{\gamma,\rho}(J)$.
For $\eta > 0$, there exists a positive constant ϵ such that $B_\eta = \{u \in PC_{\gamma,\rho}(J) : \|u\|_{PC_{\gamma,\rho}} \leq \eta\}$, we have $\|\Psi(u)\|_{PC_{\gamma,\rho}} \leq \epsilon$.
By (5.4.1) and from (5.13), we have for each $t \in I_k, k = 0, \ldots, m$,

$$\left| \left(\frac{t^\rho - s_k^\rho}{\rho} \right)^{1-\gamma} h(t) \right| = \left| \left(\frac{t^\rho - s_k^\rho}{\rho} \right)^{1-\gamma} f(t, u(t), h(t)) \right|$$

$$\leq \left(\frac{t^\rho - s_k^\rho}{\rho} \right)^{1-\gamma} (p_1(t) + p_2(t)|u(t)| + p_3(t)|h(t)|),$$

which implies that

$$\left| \left(\frac{t^\rho - s_k^\rho}{\rho} \right)^{1-\gamma} h(t) \right| \leq p_1^* \left(\frac{b^\rho - a^\rho}{\rho} \right)^{1-\gamma} + p_2^* \eta + p_3^* \left| \left(\frac{t^\rho - s_k^\rho}{\rho} \right)^{1-\gamma} h(t) \right|.$$

Then

$$\left| \left(\frac{t^\rho - s_k^\rho}{\rho} \right)^{1-\gamma} h(t) \right| \leq \frac{p_1^* \left(\frac{b^\rho - a^\rho}{\rho} \right)^{1-\gamma} + p_2^* \eta}{1 - p_3^*} := \Lambda.$$

Thus, for $t \in I_k, k = 0, \ldots, m$, (5.13) implies

$$\left| \left(\frac{t^\rho - s_k^\rho}{\rho} \right)^{1-\gamma} (\Psi u)(t) \right| \leq \frac{|\phi_k|}{\Gamma(\gamma)} + \left(\frac{t^\rho - s_k^\rho}{\rho} \right)^{1-\gamma} \left({}^\rho \mathcal{J}_{s_k^+}^\alpha |h(s)| \right)(t).$$

By Lemma 2.19, for $t \in I_k, k = 0, \ldots, m$, we have

$$\left| \left(\frac{t^\rho - s_k^\rho}{\rho} \right)^{1-\gamma} (\Psi u)(t) \right| \leq \frac{\phi^*}{\Gamma(\gamma)} + \Lambda \left(\frac{\Gamma(\gamma)}{\Gamma(\gamma + \alpha)} \right) \left(\frac{b^\rho - a^\rho}{\rho} \right)^\alpha$$

$$:= r_1.$$

And for each $t \in \tilde{I}_k, k = 1, \ldots, m$, we have

$$|\Psi u(t)|_{PC_{\gamma,\rho}} \leq |g_k(t, u(t))| \leq \Phi_1 \eta + \Phi_2 := r_2.$$

Thus, for each $t \in J$, we have

$$\|\Psi u\|_{PC_{\gamma,\rho}} \leq max\{r_1, r_2\} := \epsilon.$$

Step 3: Ψ maps bounded sets into equicontinuous sets of $PC_{\gamma,\rho}(J)$.
Let $\epsilon_1, \epsilon_2 \in J, \epsilon_1 < \epsilon_2, B_\eta$ be a bounded set of $PC_{\gamma,\rho}(J)$ as in Step 2, and let $u \in B_\eta$. Then for each $t \in I_k, k = 0, \ldots, m$, and by Lemma 2.19, we have

$$\left| \left(\frac{\epsilon_1^\rho - s_k^\rho}{\rho} \right)^{1-\gamma} (\Psi u)(\epsilon_1) - \left(\frac{\epsilon_2^\rho - s_k^\rho}{\rho} \right)^{1-\gamma} (\Psi u)(\epsilon_2) \right|$$

$$\leq \left| \left(\frac{\epsilon_1^\rho - s_k^\rho}{\rho} \right)^{1-\gamma} \left({}^\rho \mathcal{J}_{s_k^+}^\alpha h(\tau) \right)(\epsilon_1) - \left(\frac{\epsilon_2^\rho - s_k^\rho}{\rho} \right)^{1-\gamma} \left({}^\rho \mathcal{J}_{s_k^+}^\alpha h(\tau) \right)(\epsilon_2) \right|$$

$$\leq \left(\frac{\epsilon_2^\rho - s_k^\rho}{\rho} \right)^{1-\gamma} \left({}^\rho \mathcal{J}_{\epsilon_1^+}^\alpha |h(\tau)| \right)(\epsilon_2) + \frac{1}{\Gamma(\alpha)} \int_{s_k}^{\epsilon_1} \left| \tau^{\rho-1} H(\tau)h(\tau) \right| d\tau,$$

where

$$H(\tau) = \left[\left(\frac{\epsilon_1^\rho - s_k^\rho}{\rho} \right)^{1-\gamma} \left(\frac{\epsilon_1^\rho - \tau^\rho}{\rho} \right)^{\alpha-1} - \left(\frac{\epsilon_2^\rho - s_k^\rho}{\rho} \right)^{1-\gamma} \left(\frac{\epsilon_2^\rho - \tau^\rho}{\rho} \right)^{\alpha-1} \right].$$

Then by Lemma 2.19, we have

$$
\left| \left(\frac{\epsilon_1^\rho - s_k^\rho}{\rho} \right)^{1-\gamma} (\Psi u)(\epsilon_1) - \left(\frac{\epsilon_2^\rho - s_k^\rho}{\rho} \right)^{1-\gamma} (\Psi u)(\epsilon_2) \right|
$$

$$
\leq \frac{\Lambda \Gamma(\gamma)}{\Gamma(\alpha + \gamma)} \left(\frac{\epsilon_2^\rho - s_k^\rho}{\rho} \right)^{1-\gamma} \left(\frac{\epsilon_2^\rho - \epsilon_1^\rho}{\rho} \right)^{\alpha + \gamma - 1}
$$

$$
+ \Lambda \int_{s_k}^{\epsilon_1} \left| H(\tau) \frac{\tau^{\rho - 1}}{\Gamma(\alpha)} \right| \left(\frac{\tau^\rho - s_k^\rho}{\rho} \right)^{\gamma - 1} d\tau.
$$

And for each $t \in \tilde{I}_k, k = 1, \ldots, m$, we have

$$
|(\Psi u)(\epsilon_1) - (\Psi u)(\epsilon_2)| \leq |g_k(\epsilon_1, u(\epsilon_1)) - g_k(\epsilon_2, u(\epsilon_2))|.
$$

As $\epsilon_1 \to \epsilon_2$, the right-hand side of the above inequality tends to zero. From steps 1 to 3 with the Arzela-Ascoli theorem, we conclude that $\Psi : PC_{\gamma,\rho} \to PC_{\gamma,\rho}$ is continuous and completely continuous.

Step 4: A priori bound. Now it remains to show that the set

$$
G = \{ u \in PC_{\gamma,\rho} : u = \lambda^* \Psi(u) \text{ for some } 0 < \lambda^* < 1 \}
$$

is bounded. Let $u \in G$, then $u = \lambda^* \Psi(u)$ for some $0 < \lambda^* < 1$.
By (5.4.1), we have for each $t \in I_k, k = 0, \ldots, m$,

$$
\left| \left(\frac{t^\rho - s_k^\rho}{\rho} \right)^{1-\gamma} h(t) \right| = \left| \left(\frac{t^\rho - s_k^\rho}{\rho} \right)^{1-\gamma} f(t, u(t), h(t)) \right|
$$

$$
\leq \left(\frac{t^\rho - s_k^\rho}{\rho} \right)^{1-\gamma} (p_1(t) + p_2(t)|u(t)| + p_3(t)|h(t)|),
$$

which implies that

$$
\left| \left(\frac{t^\rho - s_k^\rho}{\rho} \right)^{1-\gamma} h(t) \right| \leq p_1^* \left(\frac{b^\rho - a^\rho}{\rho} \right)^{1-\gamma} + p_2^* \|u\|_{PC_{\gamma,\rho}}
$$

$$
+ p_3^* \left| \left(\frac{t^\rho - s_k^\rho}{\rho} \right)^{1-\gamma} h(t) \right|,
$$

then

$$\left| \left(\frac{t^\rho - s_k^\rho}{\rho} \right)^{1-\gamma} h(t) \right| \leq \frac{p_1^* \left(\frac{b^\rho - a^\rho}{\rho} \right)^{1-\gamma} + p_2^* \|u\|_{PC_{\gamma,\rho}}}{1 - p_3^*}.$$

This implies, by (5.13), (5.4.2), and by letting the estimation of Step 2, that for each $t \in I_k, k = 0, \ldots, m$, we have

$$\left| \left(\frac{t^\rho - s_k^\rho}{\rho} \right)^{1-\gamma} u(t) \right| \leq \frac{|\phi_k|}{\Gamma(\gamma)} + \frac{p_1^* \left(\frac{b^\rho - a^\rho}{\rho} \right)^{1-\gamma} + p_2^* \|u\|_{PC_{\gamma,\rho}}}{1 - p_3^*} \left(\frac{\Gamma(\gamma)}{\Gamma(\gamma + \alpha)} \right)$$

$$\times \left(\frac{b^\rho - a^\rho}{\rho} \right)^\alpha,$$

thus

$$\left| \left(\frac{t^\rho - s_k^\rho}{\rho} \right)^{1-\gamma} u(t) \right| \leq \frac{\phi^*}{\Gamma(\gamma)} + \left(\frac{p_1^* \Gamma(\gamma)}{(1 - p_3^*)\Gamma(\gamma + \alpha)} \right) \left(\frac{b^\rho - a^\rho}{\rho} \right)^{1-\gamma+\alpha}$$

$$+ \left(\frac{p_2^* \Gamma(\gamma)}{(1 - p_3^*)\Gamma(\gamma + \alpha)} \right) \left(\frac{b^\rho - a^\rho}{\rho} \right)^\alpha \|u\|_{PC_{\gamma,\rho}}.$$

And for each $t \in \tilde{I}_k, k = 1, \ldots, m$, we have

$$|u(t)| \leq |g_k(t, u(t))| \leq \Phi_1 \|u\|_{PC_{\gamma,\rho}} + \Phi_2.$$

Then, for each $t \in J$, we have

$$\|u\|_{PC_{\gamma,\rho}} \leq \chi_1 + \chi_2 \|u\|_{PC_{\gamma,\rho}},$$

where

$$\chi_1 = \max \left\{ \Phi_2, \frac{\phi^*}{\Gamma(\gamma)} + \left(\frac{p_1^* \Gamma(\gamma)}{(1 - p_3^*)\Gamma(\gamma + \alpha)} \right) \left(\frac{b^\rho - a^\rho}{\rho} \right)^{1-\gamma+\alpha} \right\},$$

and

$$\chi_2 = \max \left\{ \Phi_1, \left(\frac{p_2^* \Gamma(\gamma)}{(1 - p_3^*)\Gamma(\gamma + \alpha)} \right) \left(\frac{b^\rho - a^\rho}{\rho} \right)^\alpha \right\}.$$

Then by (5.15), we have

$$\|u\|_{PC_{\gamma,\rho}} \leq \frac{\chi_1}{1 - \chi_2} := R.$$

As a consequence of Theorem 2.47, and using Step 2 of the last result, we deduce that Ψ has a fixed point which is a solution of the problem (5.1)–(5.3). □

5.2.2   Nonlocal Impulsive Differential Equations

This part is concerned with a generalization of the results presented previously to nonlocal impulsive fractional differential equations. More precisely, we shall present some existence results for the following nonlocal problem:

$$\left({}^{\rho}\mathcal{D}^{\alpha,\beta}_{s_k^+} u \right)(t) = f\left(t, u(t), \left({}^{\rho}\mathcal{D}^{\alpha,\beta}_{s_k^+} u \right)(t) \right); \ t \in I_k, \ k = 0, \ldots, m, \tag{5.16}$$

$$u(t) = g_k(t, u(t)); \ t \in \tilde{I}_k, \ k = 1, \ldots, m, \tag{5.17}$$

$$\left({}^{\rho}\mathcal{J}^{1-\gamma}_{a^+} u \right)(a^+) + \vartheta(u) = \phi_0, \tag{5.18}$$

where ${}^{\rho}\mathcal{D}^{\alpha,\beta}_{s_k^+}, {}^{\rho}\mathcal{J}^{1-\gamma}_{a^+}$ are the generalized Hilfer fractional derivative of order $\alpha \in (0, 1)$ and type $\beta \in [0, 1]$ and generalized Hilfer fractional integral of order $1 - \gamma$, $(\gamma = \alpha + \beta - \alpha\beta)$, respectively, $\phi_0 \in \mathbb{R} \rho > 0$ and $I_k, \tilde{I}_k, f, g_k$ are as in the last section, and $\vartheta : PC_{\gamma,\rho}(J) \mapsto \mathbb{R}$ is a continuous function. Nonlocal conditions were initiated by Byszewski [60] when he proved the existence and uniqueness of mild and classical solutions of nonlocal Cauchy problems. The nonlocal condition can be more useful than the standard initial condition to describe some physical phenomena.

Theorem 5.5 *Assume that (5.3.1)–(5.3.3), (5.4.2), and the hypothesis that follows hold.*

(5.5.1) *There exist constants $K^* > 0$ such that*

$$|\vartheta(u) - \vartheta(\bar{u})| \le K^*|u(t) - \bar{u}(t)|$$

for any $u, \bar{u} \in PC_{\gamma,\rho}(J)$.

If

$$l^* + K^* + \frac{\Upsilon\Gamma(\gamma)}{\Gamma(\gamma + \alpha)}\left(\frac{b^\rho - a^\rho}{\rho} \right)^\alpha < 1, \tag{5.19}$$

then the nonlocal problem (5.16)–(5.18) has a unique solution in $PC_{\gamma,\rho}(J)$.

Proof We transform the problem (5.16)–(5.18) into a fixed point problem. Consider the operator $\tilde{\Psi} : PC_{\gamma,\rho}(J) \longrightarrow PC_{\gamma,\rho}(J)$ defined by

$$
\tilde{\Psi}u(t) = \begin{cases} \dfrac{\phi_0 - \vartheta(u)}{\Gamma(\gamma)} \left(\dfrac{t^\rho - a^\rho}{\rho}\right)^{\gamma-1} + \left({}^\rho \mathcal{J}_{a+}^\alpha h\right)(t) & \text{if } t \in I_0, \\[12pt] \dfrac{\phi_k}{\Gamma(\gamma)} \left(\dfrac{t^\rho - s_k^\rho}{\rho}\right)^{\gamma-1} + \left({}^\rho \mathcal{J}_{s_k^+}^\alpha h\right)(t) & \text{if } t \in I_k, k = 1, \dots, m, \\[12pt] g_k(t, u(t)) & \text{if } t \in \tilde{I}_k, k = 1, \dots, m. \end{cases} \tag{5.20}
$$

where $h \in C_{\gamma, \rho}(I_k)$, $k = 0, \dots, m$ be a function satisfying the functional equation

$$
h(t) = f(t, u(t), h(t)).
$$

Clearly, the fixed points of the operator $\tilde{\Psi}$ are solutions of the problem (5.16)–(5.18). We can easily show that $\tilde{\Psi}$ is a contraction and its fixed points are in $PC_{\gamma, \rho}^\gamma(J)$. □

Theorem 5.6 *Assume* (5.3.1), (5.4.1)–(5.4.2), *and* (5.5.1) *hold. If*

$$
max\left\{\Phi_1, \left(\frac{p_2^* \Gamma(\gamma)}{(1 - p_3^*)\Gamma(\gamma + \alpha)}\right)\left(\frac{b^\rho - a^\rho}{\rho}\right)^\alpha\right\} < 1, \tag{5.21}
$$

then the nonlocal problem (5.16)–(5.18) *has at least one solution in* $PC_{\gamma, \rho}(J)$.

5.2.3 Ulam-Hyers-Rassias Stability

First, we consider the Ulam Stability for problem (5.1)–(5.3). Let $u \in PC_{\gamma, \rho}(J)$, $\epsilon > 0$, $\tau > 0$, and $\theta : J \longrightarrow [0, \infty)$ be a continuous function. We consider the following inequalities:

$$
\begin{cases} \left|\left({}^\rho \mathcal{D}_{s_k^+}^{\alpha,\beta} u\right)(t) - f\left(t, u(t), \left({}^\rho \mathcal{D}_{s_k^+}^{\alpha,\beta} u\right)(t)\right)\right| \le \epsilon \theta(t), \ t \in I_k, \ k = 0, \dots, m, \\[12pt] |u(t) - g_k(t, u(t))| \le \epsilon \tau, \ t \in \tilde{I}_k, \ k = 1, \dots, m. \end{cases}
$$

$$\tag{5.22}$$

Definition 5.7 ([156, 158]) Problem (5.1)–(5.3) is Ulam-Hyers-Rassias (U-H-R) stable with respect to (θ, τ) if there exists a real number $a_{f, \theta} > 0$ such that for each $\epsilon > 0$ and for each solution $u \in PC_{\gamma, \rho}(J)$ of inequality (5.63) there exists a solution $w \in PC_{\gamma, \rho}(J)$ of (5.1)–(5.3) with

$$
|u(t) - w(t)| \le \epsilon a_{f, \theta}(\theta(t) + \tau), \quad t \in J.
$$

Remark 5.8 ([156, 158]) A function $u \in PC_{\gamma, \rho}(J)$ is a solution of inequality (5.22) if and only if there exist $\sigma \in PC_{\gamma, \rho}(J)$ and a sequence σ_k, $k = 0, \dots, m$ such that

1. $|\sigma(t)| \le \epsilon\theta(t), t \in I_k, k = 0, \ldots, m$; and $|\sigma_k| \le \epsilon\tau, \ t \in \tilde{I}_k, \ k = 1, \ldots, m$,

2. $\left({}^{\rho}\mathcal{D}^{\alpha,\beta}_{s^+_k} u\right)(t) = f\left(t, u(t), \left({}^{\rho}\mathcal{D}^{\alpha,\beta}_{s^+_k} u\right)(t)\right) + \sigma(t), \ t \in I_k, \ k = 0, \ldots, m$,

3. $u(t) = g_k(t, u(t)) + \sigma_k, t \in \tilde{I}_k, \ k = 1, \ldots, m$.

Theorem 5.9 *Assume that in addition to (5.3.1)–(5.3.3) and (5.14), the following hypothesis holds.*

(5.9.1) *There exist a nondecreasing function $\theta : J \longrightarrow [0, \infty)$ and $\lambda_\theta > 0$ such that for each $t \in I_k$; $k = 0, \ldots, m$, we have*

$$\left({}^{\rho}\mathcal{J}^{\alpha}_{s^+_k}\theta\right)(t) \le \lambda_\theta\theta(t).$$

Then problem (5.1)–(5.3) is U-H-R stable with respect to (θ, τ).

Proof Consider the operator Ψ is defined as in (5.13). Let $u \in PC_{\gamma,\rho}(J)$ be a solution of inequality (5.22), and let us assume that w is the unique solution of the problem

$$\begin{cases} \left({}^{\rho}\mathcal{D}^{\alpha,\beta}_{s^+_k} w\right)(t) = f\left(t, w(t), \left({}^{\rho}\mathcal{D}^{\alpha,\beta}_{s^+_k} w\right)(t)\right); \ t \in I_k, \ k = 0, \ldots, m, \\ u(t) = g_k(y, w(t^-_k)); \ t \in \tilde{I}_k, \ k = 1, \ldots, m, \\ \left({}^{\rho}\mathcal{J}^{1-\gamma}_{s^+_k} w\right)(s^+_k) = \left({}^{\rho}\mathcal{J}^{1-\gamma}_{s^+_k} u\right)(s^+_k) = \phi_k, \ k = 0, \ldots, m. \end{cases}$$

By Lemma 5.2, we obtain for each $t \in (a, b]$

$$w(t) = \begin{cases} \dfrac{\phi_k}{\Gamma(\gamma)}\left(\dfrac{t^\rho - s^\rho_k}{\rho}\right)^{\gamma-1} + \left({}^{\rho}\mathcal{J}^{\alpha}_{s^+_k} h\right)(t) & \textit{if } t \in I_k, k = 0, \ldots, m, \\ \\ g_k(t, w(t)) & \textit{if } t \in \tilde{I}_k, k = 1, \ldots, m, \end{cases}$$

where $h \in C_{\gamma,\rho}(I_k)$; $k = 0, \ldots, m$, be a function satisfying the functional equation

$$h(t) = f(t, w(t), h(t)).$$

Since u is a solution of the inequality (5.22), by Remark 5.8, we have

$$\begin{cases} \left({}^{\rho}\mathcal{D}^{\alpha,\beta}_{s^+_k} u\right)(t) = f\left(t, u(t), \left({}^{\rho}\mathcal{D}^{\alpha,\beta}_{s^+_k} u\right)(t)\right) + \sigma(t), t \in I_k, k = 0, \ldots, m; \\ u(t) = g_k(t, u(t)) + \sigma_k, t \in \tilde{I}_k, k = 1, \ldots, m. \end{cases} \tag{5.23}$$

Clearly, the solution of (5.23) is given by

$$u(t) = \begin{cases} \dfrac{\phi_k}{\Gamma(\gamma)} \left(\dfrac{t^\rho - s_k^\rho}{\rho}\right)^{\gamma-1} + \left({}^\rho J_{s_k^+}^\alpha g\right)(t) + \left({}^\rho J_{s_k^+}^\alpha \sigma\right)(t), & t \in I_k, k \neq 0, \\ g_k(t, u(t)) + \sigma_k, & t \in \tilde{I}_k, k = 1, \ldots, m, \end{cases}$$

where $g : I_k \to \mathbb{R}, k = 0, \ldots, m$, be a function satisfying the functional equation

$$g(t) = f(t, u(t), g(t)).$$

Hence, for each $t \in I_k, k = 0, \ldots, m$, we have

$$|u(t) - w(t)| \leq \left({}^\rho J_{s_k^+}^\alpha |g(s) - h(s)|\right)(t) + \left({}^\rho J_{s_k^+}^\alpha |\sigma(s)|\right)$$

$$\leq \epsilon\lambda_\theta\theta(t) + \Upsilon \int_{s_k}^t s^{\rho-1} \left(\dfrac{t^\rho - s^\rho}{\rho}\right)^{\alpha-1} \dfrac{|u(s) - w(s)|}{\Gamma(\alpha)} ds.$$

We apply Lemma 2.40 to obtain

$$|u(t) - w(t)| \leq \epsilon\lambda_\theta\theta(t) + \int_{s_k}^t \sum_{\tau=1}^\infty \dfrac{(\Upsilon)^\tau}{\Gamma(\tau\alpha)} s^{\rho-1} \left(\dfrac{t^\rho - s^\rho}{\rho}\right)^{\tau\alpha-1} (\epsilon\lambda_\theta\theta(s)) ds$$

$$\leq \epsilon\lambda_\theta\theta(t) E_\alpha\left[\Upsilon\left(\dfrac{t^\rho - s_k^\rho}{\rho}\right)^\alpha\right]$$

$$\leq \epsilon\lambda_\theta\theta(t) E_\alpha\left[\Upsilon\left(\dfrac{b^\rho - a^\rho}{\rho}\right)^\alpha\right].$$

And for each $t \in \tilde{I}_k, k = 1, \ldots, m$, we have

$$|u(t) - w(t)| \leq |g_k(t, u(t)) - g_k(t, w(t))| + |\sigma_k|$$
$$\leq l^*|u(t) - w(t)| + \epsilon\tau,$$

then by 5.14, we have

$$|u(t) - w(t)| \leq \dfrac{\epsilon\tau}{1 - l^*}.$$

Then for each $t \in J$, we have

$$|u(t) - w(t)| \leq a_\theta\epsilon(\tau + \theta(t)),$$

where

$$a_\theta = \dfrac{1}{1 - l^*} + \lambda_\theta E_\alpha\left[\Upsilon\left(\dfrac{b^\rho - a^\rho}{\rho}\right)^\alpha\right].$$

Hence, problem (5.1)–(5.3) is U-H-R stable with respect to (θ, τ). Now we are concerned with the Ulam-Hyers-Rassias stability of our problem (5.16)–(5.18). $\square$

Theorem 5.10 *Assume that (5.3.1), (5.4.1), (5.4.2), (5.5.1), (5.9.1), and (5.21) hold. Then the problem (5.16)–(5.18) is U-H-R stable with respect to (θ, τ).*

5.2.4 Examples

Example 5.11 Consider the following impulsive Cauchy problem:

$$\left(\tfrac{1}{2}\mathcal{D}^{\frac{1}{2},0}_{s_k^+}u\right)(t) = \frac{e^{-t}}{79e^{t+3}\left(1 + |u(t)| + |\tfrac{1}{2}\mathcal{D}^{\frac{1}{2},0}_{s_k^+}u(t)|\right)}, \quad \text{for each } t \in I_0 \cup I_1, \qquad (5.24)$$

$$u(t) = \frac{|u(t)|}{e^t + 2|u(t)|}, \quad \text{for each } t \in \tilde{I}_1, \qquad (5.25)$$

$$\left(\tfrac{1}{2}\mathcal{J}^{\frac{1}{2}}_{1^+}u\right)(1^+) = 0, \qquad (5.26)$$

where

$$I_0 = (1, 2], \ I_1 = (e, 3], \ \tilde{I}_1 = (2, e], \ s_0 = 1, \ t_1 = 2, \text{ and } s_1 = e.$$

Set

$$f(t, u, w) = \frac{e^{-t}}{79e^{t+3}(1 + |u| + |w|)}, \quad t \in I_0 \cup I_1, \ u, w \in \mathbb{R}.$$

We have

$$C^{\beta(1-\alpha)}_{\gamma,\rho}((1, 2]) = C^0_{\frac{1}{2},\frac{1}{2}}((1, 2])$$

$$= \left\{ v : (1, 2] \to \mathbb{R} : \sqrt{2}\left(\sqrt{t} - 1\right)^{\frac{1}{2}} v \in C([1, 2], \mathbb{R}) \right\},$$

and

$$C^{\beta(1-\alpha)}_{\gamma,\rho}((e, 3]) = C^0_{\frac{1}{2},\frac{1}{2}}((e, 3])$$

$$= \left\{ v : (e, 3] \to \mathbb{R} : \sqrt{2}\left(\sqrt{t} - \sqrt{2}\right)^{\frac{1}{2}} v \in C([e, 3], \mathbb{R}) \right\},$$

with

$$\gamma = \alpha = \frac{1}{2} \quad \rho = \frac{1}{2}, \quad \beta = 0, \text{ and } k \in \{0, 1\}.$$

Clearly, the continuous function $f \in C^0_{\frac{1}{2},\frac{1}{2}}((1, 2]) \cap C^0_{\frac{1}{2},\frac{1}{2}}((e, 3])$.
Hence, the condition (5.3.1) is satisfied.

For each $u, \bar{u}, w, \bar{w} \in \mathbb{R}$ and $t \in I_0 \cup I_1$, we have

$$|f(t, u, w) - f(t, \bar{u}, \bar{w})| \leq \frac{e^{-t}}{79e^{t+3}}(|u - \bar{u}| + |w - \bar{w}|)$$

$$\leq \frac{1}{79e^5}(|u - \bar{u}| + |w - \bar{w}|).$$

Hence condition (5.3.2) is satisfied with $\mathfrak{M}_1 = \mathfrak{M}_2 = \frac{1}{79e^5}$.

And let

$$g_1(u) = \frac{u}{e^t + 2u}, u \in [0, \infty).$$

Let $u, w \in [0, \infty)$. Then we have

$$|g_1(u) - g_1(w)| = \left| \frac{u}{e^t + 2u} - \frac{w}{e^t + 2w} \right| = \frac{e^t |u - w|}{(e^t + 2u)(e^t + 2w)} \leq \frac{1}{e}|u - w|,$$

and so the condition (5.3.3) is satisfied with $l^* = \frac{1}{e}$.

A simple computation shows that the condition (5.14) of Theorem 5.3 is satisfied, for

$$L = \frac{1}{e} + \frac{\sqrt{2\pi}(\sqrt{3} - 1)^{\frac{1}{2}}}{(79e^5 - 1)} \approx 0.368062377 < 1.$$

Then the problem (5.24)–(5.26) has a unique solution in $PC_{\frac{1}{2}, \frac{1}{2}}([1, 3])$. Also, hypothesis (5.9.1) is satisfied with $\tau = 1$ and

$$\theta(t) = \begin{cases} 2(\sqrt{t} - \sqrt{s_k}), & \text{if } t \in I_0 \cup I_1, \\ e, & \text{if } t \in \tilde{I}_1, \end{cases}$$

and $\lambda_\theta = \dfrac{\sqrt{2}\Gamma(2)(\sqrt{2} - 1)^{\frac{1}{2}}}{\Gamma(\frac{5}{2})}$. Indeed, for each $t \in I_0 \cup I_1$, we get

$$(\tfrac{1}{2}J_{1^+}^{\frac{1}{2}}\theta)(t) \leq \frac{\sqrt{2}\Gamma(2)(\sqrt{2} - 1)^{\frac{1}{2}}}{\Gamma(\frac{5}{2})}(2\sqrt{t} - 2),$$

and

$$(\tfrac{1}{2}J_{e^+}^{\frac{1}{2}}\theta)(t) \leq \frac{\sqrt{2}\Gamma(2)(\sqrt{3} - \sqrt{e})^{\frac{1}{2}}}{\Gamma(\frac{5}{2})}(2\sqrt{t} - 2\sqrt{e}).$$

Consequently, Theorem 5.9 implies that the problem (5.24)–(5.26) is U-H-R stable.

Example 5.12 Consider the following impulsive nonlocal initial value problem:

$$\left({}^{1}D_{s_k^+}^{\frac{1}{2},0}u\right)(t) = \frac{1 + |u(t)| + |{}^{1}D_{s_k^+}^{\frac{1}{2},0}u(t)|}{107e^{-t+3}(1 + |u(t)| + |{}^{1}D_{s_k^+}^{\frac{1}{2},0}u(t)|)}, \quad t \in I_k, k = 0, \ldots, 4, \quad (5.27)$$

$$u(t) = \frac{|u(t)|}{10e^k + |u(t)|}, \quad \text{for each } t \in \tilde{I}_k, k = 1, \ldots, 4, \quad (5.28)$$

$$\left({}^{1}J_{1^+}^{\frac{1}{2}}u\right)(1^+) + \frac{1}{5}\frac{u(t)}{|u(t)| + 1} = 1, \quad (5.29)$$

where

$$I_k = (s_k, t_{k+1}], \quad s_k = 1 + \frac{2k}{9} \quad \text{for } k = 0, \ldots, 4$$

and

$$\tilde{I}_k = (t_k, s_k], \quad t_k = 1 + \frac{2k-1}{9} \quad \text{for } k = 1, \ldots, 4, \ (m = 4),$$

and

$$a = s_0 = 1, \quad b = t_5 = 2.$$

Set

$$f(t, u, w) = \frac{1 + |u| + |w|}{107e^{-t+3}(1 + |u| + |w|)}, \quad t \in I_k, k = 0, \ldots, 4, \ u, w \in \mathbb{R}.$$

We have

$$C_{\gamma,\rho}^{\beta(1-\alpha)}((s_k, t_{k+1}]) = C_{\frac{1}{2},1}^{0}((s_k, t_{k+1}])$$

$$= \left\{v : (s_k, t_{k+1}] \to \mathbb{R} : (\sqrt{t - s_k})v \in C([s_k, t_{k+1}], \mathbb{R})\right\},$$

with $\gamma = \alpha = \frac{1}{2}$, $\rho = 1$, $\beta = 0$, and $k = 0, \ldots, 4$. Clearly, the continuous function $f \in C_{\frac{1}{2},1}^{0}([s_k, t_{k+1}]); k = 0, \ldots, 4$. Hence, the condition (5.3.1) is satisfied.
For each $u, w \in \mathbb{R}$ and $t \in I_k; k = 0, \ldots, 4$, we have

$$|f(t, u, w)| \le \frac{1}{107e^{-t+3}}(1 + |u| + |w|).$$

Hence, condition (5.4.1) is satisfied with

$$p_1(t) = p_2(t) = p_3(t) = \frac{1}{107e^{-t+3}},$$

and

$$p_1^* = p_2^* = p_3^* = \frac{1}{107e}.$$

Let

$$g_k(u) = \frac{u}{10e^k + u}, k = 1, \ldots, 4, u \in [0, \infty),$$

then we have

$$|g_k(u)| \le \frac{1}{10e}|u| + 1, k = 1, \ldots, 4,$$

and so the condition (5.4.2) is satisfied with $\Phi_1 = \frac{1}{10e}$ and $\Phi_2 = 1$.
And let

$$\vartheta(u) = \frac{1}{5}\frac{u}{|u| + 1}, u \in \mathbb{R}$$

then we have

$$|g(u)| \le 42 \sup\{u(t_k), k = 1, \ldots, 4\},$$

and so the condition (5.5.1) is satisfied with $\tilde{M} = 42 \sup\{u(t_k), k = 1, \ldots, 4\}$.
The condition (5.21) of Theorem 5.6 is satisfied, for

$$\Phi_1 = \frac{1}{10e} < 1,$$

and

$$\left(\frac{p_2^* \Gamma(\gamma)}{(1 - p_3^*)\Gamma(\gamma + \alpha)}\right)\left(\frac{b^\rho - a^\rho}{\rho}\right)^\alpha = \frac{\sqrt{\pi}}{(107e - 1)} < 1.$$

Then the problem (5.27)–(5.29) has at least one solution in $PC_{\frac{1}{2},1}([1, 2])$. Also, hypothesis
(5.9.1) is satisfied with $\tau = 1$ and

$$\theta(t) = \begin{cases} (t - s_k)^2, & if\ t \in I_k, k = 0, \ldots, 4, \\ 2, & if\ t \in \tilde{I}_k, k = 1, \ldots, 4, \end{cases}$$

and $\lambda_\theta = \frac{\Gamma(3)}{\Gamma(\frac{7}{2})}$. Indeed, for each $t \in I_k, k = 0, \ldots, 4$, we get

$$({}^1\mathcal{J}_{s_k^+}^{\frac{1}{2}}\theta)(t) \le \frac{\Gamma(3)}{\Gamma(\frac{7}{2})}(t - s_k)^{\frac{5}{2}}$$

$$\le \frac{\Gamma(3)}{\Gamma(\frac{7}{2})}(t - s_k)^2$$

$$= \lambda_\theta \theta(t).$$

Consequently, Theorem 5.10 implies that the problem (5.27)–(5.29) is U-H-R stable.

5.3   Initial Value Problem for Nonlinear Implicit Generalized Hilfer-Type Fractional Differential Equations in Banach Spaces

Motivated by the works mentioned in the introduction, in this section, we establish existence results to the initial value problem of nonlinear implicit generalized Hilfer-type fractional differential equation with non-instantaneous impulses:

$$\left({}^{\rho}\mathcal{D}^{\alpha,\beta}_{s_k^+} u \right)(t) = f\left(t, u(t), \left({}^{\rho}\mathcal{D}^{\alpha,\beta}_{s_k^+} u \right)(t) \right); \ t \in I_k, \ k = 0, \ldots, m, \tag{5.30}$$

$$u(t) = g_k(t, u(t)); \ t \in \tilde{I}_k, \ k = 1, \ldots, m, \tag{5.31}$$

$$\left({}^{\rho}\mathcal{J}^{1-\gamma}_{a^+} u \right)(a^+) = \phi_0, \tag{5.32}$$

where ${}^{\rho}\mathcal{D}^{\alpha,\beta}_{s_k^+}$, ${}^{\rho}\mathcal{J}^{1-\gamma}_{a^+}$ are the generalized Hilfer-type fractional derivative of order $\alpha \in (0, 1)$ and type $\beta \in [0, 1]$ and generalized fractional integral of order $1 - \gamma$, $(\gamma = \alpha + \beta - \alpha\beta)$, respectively, $\rho > 0$, $\phi_0 \in E$, $I_k := (s_k, t_{k+1}]; k = 0, \ldots, m$, $\tilde{I}_k := (t_k, s_k]; k = 1, \ldots, m$, $a = s_0 < t_1 \leq s_1 < t_2 \leq s_2 < \cdots \leq s_{m-1} < t_m \leq s_m < t_{m+1} = b < \infty$, $u(t_k^+) = \lim_{\epsilon \to 0^+} u(t_k + \epsilon)$ and $u(t_k^-) = \lim_{\epsilon \to 0^-} u(t_k + \epsilon)$ represent the right- and left-hand limits of $u(t)$ at $t = t_k$, $f : I_k \times E \times E \to E$ is a given function, and $g_k : \tilde{I}_k \times E \to E; k = 1, \ldots, m$ are given continuous functions such that $\left({}^{\rho}\mathcal{J}^{1-\gamma}_{s_k^+} g_k \right)(t, u(t)) \big|_{t=s_k} = \phi_k \in E$, where $(E, \| \cdot \|)$ is a real Banach space.

5.3.1   Existence Results

Consider the Banach space

$$PC_{\gamma,\rho}(J) = \Big\{ u : J \to E : u \in C_{\gamma,\rho}(I_k); k = 0, \ldots, m, \text{ and}$$

$$u \in C(\tilde{I}_k, E); k = 1, \ldots, m, \text{ and there exist } u(t_k^-), u(t_k^+),$$

$$u(s_k^-), \text{ and } u(s_k^+) \text{ with } u(t_k^-) = u(t_k) \Big\},$$

and

$$PC^n_{\gamma,\rho}(J) = \Big\{ u \in PC^{n-1}(J) : u^{(n)} \in PC_{\gamma,\rho}(J) \Big\}, n \in \mathbb{N},$$

$$PC^0_{\gamma,\rho}(J) = PC_{\gamma,\rho}(J),$$

with the norm

$$\|u\|_{PC_{\gamma,\rho}}$$

$$= \max \left\{ \max_{k=0,\dots,m} \left\{ \sup_{t\in[s_k,t_{k+1}]} \left\| \left(\frac{t^\rho - s_k^\rho}{\rho}\right)^{1-\gamma} u(t) \right\| \right\}, \max_{k=1,\dots,m} \left\{ \sup_{t\in[t_k,s_k]} \|u(t)\| \right\} \right\}.$$

Also, we define the following Banach space:

$$PC_{\gamma,\rho}^\gamma(J) = \left\{ u : J \to E : u \in C_{\gamma,\rho}^\gamma(I_k); k = 0, \dots, m, \text{ and} \right.$$

$$u \in C(\tilde{I}_k, E); k = 1, \dots, m, \text{ and there exist } u(t_k^-), u(t_k^+),$$

$$\left. u(s_k^-), \text{ and } u(s_k^+) \text{ with } u(t_k^-) = u(t_k) \right\}.$$

Lemma 5.13 ([75]) *Let $D \subset PC_{\gamma,\rho}(J)$ be a bounded and equicontinuous set, then
(i) the function $t \to \mu(D(t))$ is continuous on J, and*

$$\mu_{PC_{\gamma,\rho}}$$

$$= \max \left\{ \max_{k=0,\dots,m} \left\{ \sup_{t\in[s_k,t_{k+1}]} \mu\left(\left(\frac{t^\rho - s_k^\rho}{\rho}\right)^{1-\gamma} u(t) \right) \right\}, \max_{k=1,\dots,m} \left\{ \sup_{t\in[t_k,s_k]} \mu\left(u(t)\right) \right\} \right\},$$

(ii) $\mu \left(\int_a^b u(s)ds : u \in D \right) \le \int_a^b \mu(D(s))ds$, *where*

$$D(t) = \{u(t) : t \in D\}, \ t \in J.$$

Same as the last section, by following the same steps, we can have the following result:

Lemma 5.14 *Let $\gamma = \alpha + \beta - \alpha\beta$ where $0 < \alpha < 1$, $0 \le \beta \le 1$, and $k = 0, \dots, m$, let $f : I_k \times E \times E \to E$, be a function such that $f(\cdot, u(\cdot), w(\cdot)) \in C_{\gamma,\rho}(I_k)$, for any $u, w \in PC_{\gamma,\rho}(J)$. If $u \in PC_{\gamma,\rho}^\gamma(J)$, then u satisfies the problem (5.30)–(5.32) if and only if u is the fixed point of the operator $\Psi : PC_{\gamma,\rho}(J) \to PC_{\gamma,\rho}(J)$ defined by*

$$\Psi u(t) = \begin{cases} \dfrac{\phi_k}{\Gamma(\gamma)} \left(\dfrac{t^\rho - s_k^\rho}{\rho}\right)^{\gamma-1} + \left({}^\rho\mathcal{J}_{s_k^+}^\alpha h\right)(t), & t \in I_k, \ k = 0, \dots, m, \\[4mm] g_k(t, u(t)), & t \in \tilde{I}_k, \ k = 1, \dots, m, \end{cases}$$

$$(5.33)$$

where $\phi^* = \max\{\|\phi_k\| : k = 0, \dots, m\}$ and $h \in C_{\gamma,\rho}(I_k)$, $k = 0, \dots, m$ be a function satisfying the functional equation

$$h(t) = f(t, u(t), h(t)).$$

Also, by Lemma 2.23, $\Psi u \in PC_{\gamma,\rho}(J)$.

We are now in a position to state and prove our existence result for the problem (5.30)–(5.32) based on Mönch's fixed point theorem.

Theorem 5.15 *Assume that the following hypotheses are met.*

(5.15.1) *The function $t \mapsto f(t, u, w)$ is measurable on $I_k; k = 0, \ldots, m$, for each $u, w \in E$, and the functions $u \mapsto f(t, u, w)$ and $w \mapsto f(t, u, w)$ are continuous on E for a.e. $t \in I_k; k = 0, \ldots, m$, and*

$$f(\cdot, u(\cdot), w(\cdot)) \in C^{\beta(1-\alpha)}_{\gamma,\rho}(I_k) \text{ for any } u, w \in PC_{\gamma,\rho}(J).$$

(5.15.2) *There exists a continuous function $p : [a, b] \longrightarrow [0, \infty)$ such that*

$$\|f(t, u, w)\| \le p(t), \text{ for a.e. } t \in I_k; k = 0, \ldots, m, \text{ and for each } u, w \in E.$$

(5.15.3) *For each bounded set $B \subset E$ and for each $t \in I_k; k = 0, \ldots, m$, we have*

$$\mu(f(t, B, ({}^\rho \mathcal{D}^{\alpha,\beta}_{s_k^+} B))) \le p(t)\mu(B),$$

where ${}^\rho \mathcal{D}^{\alpha,\beta}_{s_k^+} B = \{{}^\rho \mathcal{D}^{\alpha,\beta}_{s_k^+} w : w \in B\}$ and $p^* = \sup\limits_{t \in [a,b]} p(t)$.

(5.15.4) *The functions $g_k \in C(\bar{I}_k, E); k = 1, \ldots, m$, and there exists $l^* > 0$ such that*

$$\|g_k(t, u)\| \le l^* \|u\| \text{ for each } u \in E, k = 1, \ldots, m.$$

(5.15.5) *For each bounded set $B \subset E$ and for each $t \in \tilde{I}_k; k = 1, \ldots, m$, we have*

$$\mu(g_k(t, B)) \le l^* \mu(B), k = 1, \ldots, m.$$

If

$$L := max \left\{ l^*, \frac{p^* \Gamma(\gamma)}{\Gamma(\alpha + \gamma)} \left(\frac{b^\rho - a^\rho}{\rho} \right)^\alpha \right\} < 1, \tag{5.34}$$

then the problem (5.30)–(5.32) has at least one solution in $PC_{\gamma,\rho}(J)$.

Proof Consider the operator $\Psi : PC_{\gamma,\rho}(J) \to PC_{\gamma,\rho}(J)$ defined in (5.33) and the ball $B_R := B(0, R) = \{w \in PC_{\gamma,\rho}(J) : \|w\|_{PC_{\gamma,\rho}} \le R\}$, such that

$$R \geq \frac{\phi^*}{(1 - l^*)\Gamma(\gamma)} + \frac{p^*}{(1 - l^*)\Gamma(\alpha + 1)} \left(\frac{b^\rho - a^\rho}{\rho}\right)^{1 - \gamma + \alpha}.$$

For any $u \in B_R$, and each $t \in I_k, k = 0, \ldots, m$, we have

$$\|\Psi u(t)\| \leq \frac{\|\phi_k\|}{\Gamma(\gamma)} \left(\frac{t^\rho - s_k^\rho}{\rho}\right)^{\gamma - 1} + \left(\rho \mathcal{J}_{s_k^+}^\alpha \|h(s)\|\right)(t)$$

$$\leq \frac{\phi^*}{\Gamma(\gamma)} \left(\frac{t^\rho - s_k^\rho}{\rho}\right)^{\gamma - 1} + p^* \left(\rho \mathcal{J}_{s_k^+}^\alpha (1)\right)(t).$$

By Lemma 2.19, we have

$$\left\|\left(\frac{t^\rho - s_k^\rho}{\rho}\right)^{1 - \gamma} \Psi u(t)\right\| \leq \frac{\phi^*}{\Gamma(\gamma)} + \frac{p^*}{\Gamma(\alpha + 1)} \left(\frac{t^\rho - s_k^\rho}{\rho}\right)^{1 - \gamma + \alpha}$$

$$\leq \frac{\phi^*}{\Gamma(\gamma)} + \frac{p^*}{\Gamma(\alpha + 1)} \left(\frac{b^\rho - a^\rho}{\rho}\right)^{1 - \gamma + \alpha}.$$

And for $t \in \tilde{I}_k, k = 1, \ldots, m$, we have

$$\|(\Psi u)(t)\| \leq l^* \|u(t)\| \leq l^* R.$$

Hence,

$$\|\Psi u\|_{PC_{\gamma, \rho}} \leq l^* R + \frac{\phi^*}{\Gamma(\gamma)} + \frac{p^*}{\Gamma(\alpha + 1)} \left(\frac{b^\rho - a^\rho}{\rho}\right)^{1 - \gamma + \alpha} \leq R.$$

This proves that Ψ transforms the ball B_R into itself. We shall show that the operator $\Psi : B_R \to B_R$ satisfies all the assumptions of Theorem 2.49. The proof will be given in several steps.

Step 1: $\Psi : B_R \to B_R$ is continuous. Let $\{u_n\}$ be a sequence such that $u_n \to u$ in $PC_{\gamma, \rho}(J)$. Then for each $t \in I_k, k = 0, \ldots, m$, we have

$$\left\|((\Psi u_n)(t) - (\Psi u)(t)) \left(\frac{t^\rho - s_k^\rho}{\rho}\right)^{1 - \gamma}\right\| \leq \left(\frac{t^\rho - s_k^\rho}{\rho}\right)^{1 - \gamma} \left(\rho \mathcal{J}_{s_k^+}^\alpha \|h_n(s) - h(s)\|\right)(t),$$

where $h_n, h \in C_{\gamma, \rho}(I_k); k = 0, \ldots, m$, such that

$$h_n(t) = f(t, u_n(t), h_n(t)),$$
$$h(t) = f(t, u(t), h(t)).$$

For each $t \in \tilde{I}_k, k = 1, \ldots, m$, we have

$$\|((\Psi u_n)(t) - (\Psi u)(t))\| \leq \|(g_k(t, u_n(t)) - g_k(t, u(t)))\|.$$

Since $u_n \to u$, then we get $h_n(t) \to h(t)$ as $n \to \infty$ for each $t \in J$, and since f and g_k are continuous, then we have

$$\|\Psi u_n - \Psi u\|_{PC_{\gamma,\rho}} \to 0 \text{ as } n \to \infty.$$

Step 2: $\Psi(B_R)$ is bounded and equicontinuous.
Since $\Psi(B_R) \subset B_R$ and B_R is bounded, then $\Psi(B_R)$ is bounded.
Next, let $\epsilon_1, \epsilon_2 \in I_k, k = 0, \ldots, m, \epsilon_1 < \epsilon_2$, and let $u \in B_R$. Then

$$\left\| \left(\frac{\epsilon_1^\rho - s_k^\rho}{\rho}\right)^{1-\gamma} (\Psi u)(\epsilon_1) - \left(\frac{\epsilon_2^\rho - s_k^\rho}{\rho}\right)^{1-\gamma} (\Psi u)(\epsilon_2) \right\|$$

$$\leq \left\| \left(\frac{\epsilon_1^\rho - s_k^\rho}{\rho}\right)^{1-\gamma} \left({}^\rho \mathcal{J}_{s_k^+}^\alpha h(\tau)\right)(\epsilon_1) - \left(\frac{\epsilon_2^\rho - s_k^\rho}{\rho}\right)^{1-\gamma} \left({}^\rho \mathcal{J}_{s_k^+}^\alpha h(\tau)\right)(\epsilon_2) \right\|$$

$$\leq \left(\frac{\epsilon_2^\rho - s_k^\rho}{\rho}\right)^{1-\gamma} \left({}^\rho \mathcal{J}_{\epsilon_1}^\alpha \|h(\tau)\|\right)(\epsilon_2) + \frac{1}{\Gamma(\alpha)} \int_{s_k}^{\epsilon_1} \left\| \tau^{\rho-1} H(\tau) h(\tau) \right\| d\tau,$$

where

$$H(\tau) = \left[\left(\frac{\epsilon_1^\rho - s_k^\rho}{\rho}\right)^{1-\gamma} \left(\frac{\epsilon_1^\rho - \tau^\rho}{\rho}\right)^{\alpha-1} - \left(\frac{\epsilon_2^\rho - s_k^\rho}{\rho}\right)^{1-\gamma} \left(\frac{\epsilon_2^\rho - \tau^\rho}{\rho}\right)^{\alpha-1} \right].$$

Then by Lemma 2.19, we have

$$\left\| \left(\frac{\epsilon_1^\rho - s_k^\rho}{\rho}\right)^{1-\gamma} (\Psi u)(\epsilon_1) - \left(\frac{\epsilon_2^\rho - s_k^\rho}{\rho}\right)^{1-\gamma} (\Psi u)(\epsilon_2) \right\|$$

$$\leq \frac{p^*}{\Gamma(1+\alpha)} \left(\frac{\epsilon_2^\rho - s_k^\rho}{\rho}\right)^{1-\gamma} \left(\frac{\epsilon_2^\rho - \epsilon_1^\rho}{\rho}\right)^\alpha$$

$$+ p^* \int_{s_k}^{\epsilon_1} \left\| H(\tau) \frac{\tau^{\rho-1}}{\Gamma(\alpha)} \right\| \left(\frac{\tau^\rho - s_k^\rho}{\rho}\right)^{\gamma-1} d\tau,$$

and for each $t \in \tilde{I}_k, k = 1, \ldots, m$, we have

$$\|(\Psi u)(\epsilon_1) - (\Psi u)(\epsilon_2)\| \leq \|(g_k(\epsilon_1, u(\epsilon_1))) - (g_k(\epsilon_2, u(\epsilon_2)))\|.$$

As $\epsilon_1 \to \epsilon_2$, the right-hand side of the above inequality tends to zero. Hence, $\Psi(B_R)$ is bounded and equicontinuous.

Step 3: The implication (2.11) of Theorem 2.49 holds.
Now let D be an equicontinuous subset of B_R such that $D \subset \overline{\Psi(D)} \cup \{0\}$; therefore, the

function $t \longrightarrow d(t) = \mu(D(t))$ is continuous on J. By (5.15.3), (5.15.5), and the properties of the measure μ, for each $t \in I_k, k = 0, \ldots, m$, we have

$$
\left(\frac{t^{\rho} - s_k^{\rho}}{\rho}\right)^{1-\gamma} d(t) \leq \mu\left(\left(\frac{t^{\rho} - s_k^{\rho}}{\rho}\right)^{1-\gamma} (\Psi D)(t) \cup \{0\}\right)
$$

$$
\leq \mu\left(\left(\frac{t^{\rho} - s_k^{\rho}}{\rho}\right)^{1-\gamma} (\Psi D)(t)\right),
$$

then

$$
\left(\frac{t^{\rho} - s_k^{\rho}}{\rho}\right)^{1-\gamma} d(t) \leq \left(\frac{t^{\rho} - s_k^{\rho}}{\rho}\right)^{1-\gamma} \left({}^{\rho}\mathcal{J}_{s_k^+}^{\alpha} p(s)\mu(D(s))\right)(t)
$$

$$
\leq p^* \left(\frac{b^{\rho} - a^{\rho}}{\rho}\right)^{1-\gamma} \left({}^{\rho}\mathcal{J}_{s_k^+}^{\alpha} d(s)\right)(t)
$$

$$
\leq \left[\frac{p^*\Gamma(\gamma)}{\Gamma(\alpha + \gamma)} \left(\frac{b^{\rho} - a^{\rho}}{\rho}\right)^{\alpha}\right] \|d\|_{PC_{\gamma,\rho}}.
$$

And for each $t \in \tilde{I}_k, k = 1, \ldots, m$, we have

$$
d(t) \leq \mu\left(g_k(t, D(t))\right) \leq l^* d(t).
$$

Thus, for each $t \in J$, we have

$$
\|d\|_{PC_{\gamma,\rho}} \leq L\|d\|_{PC_{\gamma,\rho}}.
$$

From (5.34), we get $\|d\|_{PC_{\gamma,\rho}} = 0$, that is $d(t) = \mu(D(t)) = 0$, for each $t \in J$, and then $D(t)$ is relatively compact in E. In view of the Ascoli-Arzela Theorem, D is relatively compact in B_R. Applying now Theorem 2.49, we conclude that Ψ has a fixed point $u^* \in PC_{\gamma,\rho}(J)$, which is solution of the problem (5.30)–(5.32).

Step 4: We show that such a fixed point $u^* \in PC_{\gamma,\rho}(J)$ is actually in $PC_{\gamma,\rho}^{\gamma}(J)$.
Since u^* is the unique fixed point of operator Ψ in $PC_{\gamma,\rho}(J)$, then for each $t \in J$, we have

$$
\Psi u^*(t) = \begin{cases} \dfrac{\phi_k}{\Gamma(\gamma)} \left(\dfrac{t^{\rho} - s_k^{\rho}}{\rho}\right)^{\gamma-1} + \left({}^{\rho}\mathcal{J}_{s_k^+}^{\alpha} h\right)(t), & t \in I_k, \ k = 0, \ldots, m, \\ g_k(t, u^*(t)), & t \in \tilde{I}_k, \ k = 1, \ldots, m. \end{cases}
$$

where $h \in C_{\gamma,\rho}(I_k); k = 0, \ldots, m$, such that

$$
h(t) = f(t, u^*(t), h(t)).
$$

For $t \in I_k; k = 0, \ldots, m$, applying ${}^{\rho}\mathcal{D}_{s_k^+}^{\gamma}$ to both sides and by Lemmas 2.19 and 2.33, we have

$$\begin{aligned}
{}^{\rho}\mathcal{D}_{s_k^+}^{\gamma} u^*(t) &= \left({}^{\rho}\mathcal{D}_{s_k^+}^{\gamma}\, {}^{\rho}\mathcal{J}_{s_k^+}^{\alpha} f(s, u^*(s), h(s)) \right)(t) \\
&= \left({}^{\rho}\mathcal{D}_{s_k^+}^{\beta(1-\alpha)} f(s, u^*(s), h(s)) \right)(t).
\end{aligned}$$

Since $\gamma \geq \alpha$, by condition (5.15.1), the right-hand side is in $C_{\gamma,\rho}(I_k)$ and thus ${}^{\rho}\mathcal{D}_{s_k^+}^{\gamma} u^* \in C_{\gamma,\rho}(I_k)$ which implies that $u^* \in C_{\gamma,\rho}^{\gamma}(I_k)$. And since $g_k \in C(\tilde{I}_k, E); k = 1, \ldots, m$, then $u^* \in PC_{\gamma,\rho}^{\gamma}(J)$. As a consequence of Steps 1 to 4 together with Theorem 5.15, we can conclude that the problem (5.30)–(5.32) has at least one solution in $PC_{\gamma,\rho}(J)$. $\square$

Our second existence result for the problem (5.30)–(5.32) is based on Darbo's fixed point Theorem.

Theorem 5.16 *Assume that conditions (5.15.1)–(5.15.5) hold. If*

$$L := max \left\{ l^*, \frac{p^* \Gamma(\gamma)}{\Gamma(\alpha + \gamma)} \left(\frac{b^\rho - a^\rho}{\rho} \right)^\alpha \right\} < 1,$$

then the problem (5.30)–(5.32) has at least one solution in $PC_{\gamma,\rho}(J)$.

Proof Consider the operator Ψ defined in (5.33). We know that $\Psi : B_R \longrightarrow B_R$ is bounded and continuous and that $\Psi(B_R)$ is equicontinuous, we need to prove that the operator Ψ is a L-contraction.
Let $D \subset B_R$ and $t \in I_k, k = 0, \ldots, m$. Then we have

$$\begin{aligned}
\mu \left(\left(\frac{t^\rho - s_k^\rho}{\rho} \right)^{1-\gamma} (\Psi D)(t) \right) &= \mu \left(\left(\frac{t^\rho - s_k^\rho}{\rho} \right)^{1-\gamma} (\Psi u)(t) : u \in D \right) \\
&\leq \left(\frac{b^\rho - a^\rho}{\rho} \right)^{1-\gamma} \left\{ \left({}^{\rho}\mathcal{J}_{s_k^+}^{\alpha} p^* \mu(u(s)) \right)(t), u \in D \right\}.
\end{aligned}$$

By Lemma 2.19, we have for $t \in I_k, k = 0, \ldots, m$,

$$\mu \left(\left(\frac{t^\rho - s_k^\rho}{\rho} \right)^{1-\gamma} (\Psi D)(t) \right) \leq \left[\frac{p^* \Gamma(\gamma)}{\Gamma(\alpha + \gamma)} \left(\frac{b^\rho - a^\rho}{\rho} \right)^\alpha \right] \mu_{PC_{\gamma,\rho}}(D).$$

And for each $t \in \tilde{I}_k, k = 1, \ldots, m$, we have

$$\mu ((\Psi D)(t)) \leq \mu (g_k(t, D(t))) \leq l^* \mu (D(t)).$$

Hence, for each $t \in J$, we have

$$\mu_{PC_{\gamma,\rho}}(\Psi D) \leq L \mu_{PC_{\gamma,\rho}}(D).$$

So, by (5.34), the operator Ψ is a L-contraction. As a consequence of Theorem 2.48 and using Step 4 of the last result, we deduce that Ψ has a fixed point which is a solution of the problem (5.30)–(5.32).   □

5.3.2 Ulam-Hyers-Rassias Stability

We are concerned with the Ulam-Hyers-Rassias stability of our problem (5.30)–(5.32). Let $u \in PC_{\gamma,\rho}(J), \epsilon > 0, \tau > 0$, and $\theta : J \longrightarrow [0, \infty)$ be a continuous function. We consider the following inequality:

$$\begin{cases} \left\| \left({}^\rho\mathcal{D}^{\alpha,\beta}_{s^+_k} u \right)(t) - f\left(t, u(t), \left({}^\rho\mathcal{D}^{\alpha,\beta}_{s^+_k} u \right)(t) \right) \right\| \leq \epsilon\theta(t),\ t \in I_k,\ k = 0, \ldots, m, \\ \|u(t) - g_k(t, u(t))\| \leq \epsilon\tau,\ t \in \tilde{I}_k,\ k = 1, \ldots, m. \end{cases}$$
(5.35)

Definition 5.17 ([156, 158]) Problem (5.30)–(5.32) is Ulam-Hyers-Rassias (U-H-R) stable with respect to (θ, τ) if there exists a real number $a_{f,\theta} > 0$ such that for each $\epsilon > 0$ and for each solution $u \in PC_{\gamma,\rho}(J)$ of inequality (5.35) there exists a solution $w \in PC_{\gamma,\rho}(J)$ of (5.30)–(5.32) with

$$\|u(t) - w(t)\| \leq \epsilon a_{f,\theta}(\theta(t) + \tau), \quad t \in J.$$

Remark 5.18 ([156, 158]) A function $u \in PC_{\gamma,\rho}(J)$ is a solution of inequality (5.35) if and only if there exist $\sigma \in PC_{\gamma,\rho}(J)$ and a sequence $\sigma_k,\ k = 0, \ldots, m$ such that

1. $\|\sigma(t)\| \leq \epsilon\theta(t), t \in I_k,\ k = 0, \ldots, m$; and $\|\sigma_k\| \leq \epsilon\tau, t \in \tilde{I}_k,\ k = 1, \ldots, m,$
2. $\left({}^\rho\mathcal{D}^{\alpha,\beta}_{s^+_k} u \right)(t) = f\left(t, u(t), \left({}^\rho\mathcal{D}^{\alpha,\beta}_{s^+_k} u \right)(t) \right) + \sigma(t), t \in I_k,\ k = 0, \ldots, m,$
3. $u(t) = g_k(t, u(t)) + \sigma_k, t \in \tilde{I}_k,\ k = 1, \ldots, m.$

Theorem 5.19 *Assume that in addition to (5.15.1)–(5.15.5) and (5.34), the following hypothesis holds.*

(5.19.1) *There exist a nondecreasing function $\theta : J \longrightarrow [0, \infty)$ and $\lambda_\theta > 0$ such that for each $t \in I_k; k = 0, \ldots, m$, we have*

$$({}^\rho\mathcal{J}^\alpha_{s^+_k}\theta)(t) \leq \lambda_\theta\theta(t).$$

(5.19.2) *There exists a continuous function $\chi : \bigcup_{k=1}^{m}[s_k, t_{k+1}] \longrightarrow [0, \infty)$ such that for each $t \in I_k; k = 0, \ldots, m$, we have*

$$p(t) \le \chi(t)\theta(t).$$

Then problem (5.30)–(5.32) is U-H-R stable with respect to (θ, τ).

Proof Consider the operator Ψ defined in (5.33). Let $u \in PC_{\gamma,\rho}(J)$ be a solution of inequality (5.35), and let us assume that w is the unique solution of the problem

$$\begin{cases} \left({}^{\rho}D_{s_k^+}^{\alpha,\beta} w\right)(t) = f\left(t, w(t), \left({}^{\rho}D_{s_k^+}^{\alpha,\beta} w\right)(t)\right); \ t \in I_k, \ k = 0, \dots, m, \\ w(t) = g_k(t, w(t_k^-)); \ t \in \tilde{I}_k, \ k = 1, \dots, m, \\ \left({}^{\rho}\mathcal{J}_{s_k^+}^{1-\gamma} w\right)(s_k^+) = \left({}^{\rho}\mathcal{J}_{s_k^+}^{1-\gamma} u\right)(s_k^+) = \phi_k, \ k = 0, \dots, m. \end{cases}$$

By Lemma 5.14, we obtain for each $t \in (a, b]$

$$w(t) = \begin{cases} \dfrac{\phi_k}{\Gamma(\gamma)} \left(\dfrac{t^\rho - s_k^\rho}{\rho}\right)^{\gamma-1} + \left({}^{\rho}\mathcal{J}_{s_k^+}^{\alpha} h\right)(t) & if \ t \in I_k, k = 0, \dots, m, \\ \\ g_k(t, w(t)) & if \ t \in \tilde{I}_k, k = 1, \dots, m, \end{cases}$$

where $h \in C_{\gamma,\rho}(I_k)$; $k = 0, \dots, m$, be a function satisfying the functional equation

$$h(t) = f(t, w(t), h(t)).$$

Since u is a solution of the inequality (5.35), by Remark 5.18, we have

$$\begin{cases} \left({}^{\rho}D_{s_k^+}^{\alpha,\beta} u\right)(t) = f\left(t, u(t), \left({}^{\rho}D_{s_k^+}^{\alpha,\beta} u\right)(t)\right) + \sigma(t), t \in I_k, k = 0, \dots, m; \\ u(t) = g_k(t, u(t)) + \sigma_k, t \in \tilde{I}_k, k = 1, \dots, m. \end{cases} \quad (5.36)$$

Clearly, the solution of (5.36) is given by

$$u(t) = \begin{cases} \dfrac{\phi_k}{\Gamma(\gamma)} \left(\dfrac{t^\rho - s_k^\rho}{\rho}\right)^{\gamma-1} + \left({}^{\rho}\mathcal{J}_{s_k^+}^{\alpha} g\right)(t) \\ + \left({}^{\rho}\mathcal{J}_{s_k^+}^{\alpha} \sigma\right)(t), \ t \in I_k, k = 0, \dots, m, \\ \\ g_k(t, u(t)) + \sigma_k, \quad t \in \tilde{I}_k, \ k = 1, \dots, m, \end{cases}$$

where $g : I_k \to E, k = 0, \dots, m$, be a function satisfying the functional equation

$$g(t) = f(t, u(t), g(t)).$$

Hence, for each $t \in I_k, k = 0, \dots, m$, we have

$$\|u(t) - w(t)\| \leq \left({}^{\rho}\mathcal{J}_{a+}^{\alpha} \|g(s) - h(s)\|\right)(t) + \left({}^{\rho}\mathcal{J}_{a+}^{\alpha} \|\sigma(s)\|\right)$$

$$\leq \epsilon \lambda_{\theta} \theta(t) + \int_{a}^{t} s^{\rho-1} \left(\frac{t^{\rho} - s^{\rho}}{\rho}\right)^{\alpha-1} \frac{2\chi(t)\theta(t)}{\Gamma(\gamma)} ds$$

$$\leq \epsilon \lambda_{\theta} \theta(t) + 2\chi^{*} \left({}^{\rho}\mathcal{J}_{a+}^{\alpha} \theta\right)(t)$$

$$\leq (\epsilon + 2\chi^{*})\lambda_{\theta} \theta(t)$$

$$\leq (1 + \frac{2\chi^{*}}{\epsilon})\lambda_{\theta} \epsilon(\tau + \theta(t)),$$

where

$$\chi^{*} = \max_{k=0,\dots,m} \left\{ \sup_{t \in [s_{k}, t_{k+1}]} \chi(t) \right\}.$$

For each $t \in \tilde{I}_{k}, k = 1, \dots, m$, we have

$$\|u(t) - w(t)\| \leq \|g_{k}(t, u(t)) - g_{k}(t, w(t))\| + \|\sigma_{k}\|$$

$$\leq l^{*}\|u(t) - w(t)\| + \epsilon\tau,$$

then by (5.34),

$$\|u(t) - w(t)\| \leq \frac{\epsilon\tau}{1 - l^{*}} \leq \frac{\epsilon}{1 - l^{*}}(\tau + \theta(t)).$$

Then for each $t \in (a, b]$, we have

$$\|u(t) - w(t)\| \leq a_{\theta} \epsilon(\tau + \theta(t)),$$

where

$$a_{\theta} = \max \left\{ (1 + \frac{2\chi^{*}}{\epsilon})\lambda_{\theta}, \frac{1}{1 - l^{*}} \right\}.$$

Hence, problem (5.30)–(5.32) is U-H-R stable with respect to (θ, τ).   □

5.3.3   An Example

Example 5.20 Let

$$E = l^{1} = \left\{ v = (v_{1}, v_{2}, \dots, v_{n}, \dots), \sum_{n=1}^{\infty} |v_{n}| < \infty \right\}$$

be the Banach space with the norm

$$\|v\| = \sum_{n=1}^{\infty} |v_{n}|.$$

Consider the following initial value problem with non-instantaneous impulses:

$$\left({}^1D_{s_k^+}^{\frac{1}{2},0} u \right)(t) = f\left(t, u(t), \left({}^1D_{s_k^+}^{\frac{1}{2},0} u \right)(t) \right), \; t \in (1,2] \cup (e,3], \; k \in \{0,1\} \quad (5.37)$$

$$u(t) = g(t, u(t)), \; t \in (2,e], \quad (5.38)$$

$$\left({}^1J_{1^+}^{\frac{1}{2}} u \right)(1^+) = 0, \quad (5.39)$$

where

$$a = t_0 = s_0 = 1 < t_1 = 2 < s_1 = e < t_2 = 3 = b,$$

$$u = (u_1, u_2, \ldots, u_n, \ldots),$$

$$f = (f_1, f_2, \ldots, f_n, \ldots),$$

$${}^1D_{s_k^+}^{\frac{1}{2},0} u = ({}^1D_{s_k^+}^{\frac{1}{2},0} u_1, \ldots, {}^1D_{s_k^+}^{\frac{1}{2},0} u_2, \ldots, {}^1D_{s_k^+}^{\frac{1}{2},0} u_n, \ldots),$$

$$g = (g_1, g_2, \ldots, g_n, \ldots),$$

$$f_n(t, u_n(t), \left({}^1D_{s_k^+}^{\frac{1}{2},0} u_n \right)(t)) = \frac{(2t^3 + 5e^{-2})|u_n(t)|}{183e^{-t+3}(1 + \|u(t)\| + \| \left({}^1D_{s_k^+}^{\frac{1}{2},0} u \right)(t) \|)},$$

for $t \in (1,2] \cup (e,3]$, with $k \in \{0,1\}$, $n \in \mathbb{N}$, and

$$g_n(t, u_n(t)) = \frac{|u_n(t)|}{105e^{-t+5}+1}, \; t \in (2,e], n \in \mathbb{N}.$$

We have

$$C_{\gamma,\rho}^{\beta(1-\alpha)}((1,2]) = C_{\frac{1}{2},1}^0((1,2]) = \left\{ h : (1,2] \to E : (\sqrt{t-1})h \in C([1,2], E) \right\},$$

and

$$C_{\gamma,\rho}^{\beta(1-\alpha)}((e,3]) = C_{\frac{1}{2},1}^0((e,3]) = \left\{ h : (e,3] \to E : (\sqrt{t-e})h \in C([e,3], E) \right\},$$

with $\gamma = \alpha = \frac{1}{2}$, $\rho = 1$, $\beta = 0$, and $k \in \{0,1\}$. Clearly, the continuous function $f \in C_{\frac{1}{2},1}^0((1,2]) \cap C_{\frac{1}{2},1}^0((e,3])$. Hence, the condition (5.15.1) is satisfied.
For each $u, w \in E$ and $t \in (1,2] \cup (e,3]$:

$$\|f(t,u,w)\| \le \frac{2t^3 + 5e^{-2}}{183e^{-t+3}}.$$

Hence, condition (5.15.2) is satisfied with

$$p(t) = \frac{2t^3 + 5e^{-2}}{183e^{-t+3}},$$

and

$$p^* = \frac{54 + 5e^{-2}}{183}.$$

And for each $u \in E$ and $t \in (2, e]$, we have

$$\|g(t, u)\| \leq \frac{\|u\|}{105e^{5-e} + 1},$$

and so the condition (5.15.4) is satisfied with $l^* = \dfrac{1}{105e^{5-e} + 1}$.
The condition (5.34) of Theorem 5.15 is satisfied, for

$$L := \max \left\{ l^*, \frac{p^* \Gamma(\gamma)}{\Gamma(\alpha + \gamma)} \left(\frac{b^\rho - a^\rho}{\rho} \right)^\alpha \right\} \approx 0.7489295248 < 1.$$

Let Ω be a bounded set in E where ${}^1\mathcal{D}_{s_k^+}^{\frac{1}{2},0}\Omega = \left\{ {}^1\mathcal{D}_{s_k^+}^{\frac{1}{2},0} v : v \in \Omega \right\}$; $k \in \{0, 1\}$, then by the
properties of the Kuratowski measure of noncompactness, for each $u \in \Omega$ and $t \in (1, 2] \cup (e, 3]$, we have

$$\mu \left(f(t, \Omega, {}^1\mathcal{D}_{s_k^+}^{\frac{1}{2},0}\Omega) \right) \leq p(t) \mu(\Omega),$$

and for each $t \in (2, e]$,

$$\mu(g(t, \Omega)) \leq l^* \mu(\Omega).$$

Hence, conditions (5.15.3) and (5.15.5) are satisfied. Then the problem (5.37)–(5.39) has
at least one solution in $PC_{\frac{1}{2}, 1}([1, 3])$.
Also, hypothesis (5.19.1) is satisfied with $\tau = 1$ and

$$\theta(t) = \begin{cases} \dfrac{1}{\sqrt{t - s_k}}, & \text{if } t \in (1, 2] \cup (e, 3], \\[2mm] 1, & \text{if } t \in (2, e], \end{cases}$$

and $\lambda_\theta = \sqrt{\pi}$. Indeed, for each $t \in (1, 2]$, we get

$$({}^1\mathcal{J}_{1+}^{\frac{1}{2}}\theta)(t) = \sqrt{\pi} \leq \frac{\sqrt{\pi}}{\sqrt{t - 1}},$$

and for each $t \in (e, 3]$, we get

$$({}^1\mathcal{J}_{e+}^{\frac{1}{2}}\theta)(t) = \sqrt{\pi} \leq \frac{\sqrt{\pi}}{\sqrt{t - e}}.$$

Let the function $\chi : [1, 2] \cup [e, 3] \longrightarrow [0, \infty)$ be defined by:

$$\chi(t) = \frac{(2t^3 + 5e^{-2})\sqrt{t - s_k}}{183e^{-t+3}}; k \in \{0, 1\},$$

then, for each $t \in (1, 2] \cup (e, 3]$, we have

$$p(t) = \chi(t)\theta(t),$$

with $\chi^* = p^*$. Hence, the condition (5.19.2) is satisfied. Consequently, Theorem 5.19 implies that the problem (5.37)–(5.39) is U-H-R stable.

5.4   Boundary Value Problem for Fractional Order Generalized Hilfer-Type Fractional Derivative

Following the work of the previous section, in this section, we establish existence and stability results of the boundary value problem with nonlinear implicit generalized Hilfer-type fractional differential equation with non-instantaneous impulses:

$$\left({}^{\rho}\mathcal{D}^{\alpha,\beta}_{\tau_i^+} x \right)(t) = f\left(t, x(t), \left({}^{\rho}\mathcal{D}^{\alpha,\beta}_{\tau_i^+} x \right)(t) \right); \ t \in J_i, \ i = 0, \ldots, m, \tag{5.40}$$

$$x(t) = \psi_i(t, x(t)); \ t \in \tilde{J}_i, \ i = 1, \ldots, m, \tag{5.41}$$

$$\phi_1 \left({}^{\rho}\mathcal{J}^{1-\gamma}_{a^+} x \right)(a^+) + \phi_2 \left({}^{\rho}\mathcal{J}^{1-\gamma}_{m^+} x \right)(b) = \phi_3, \tag{5.42}$$

where ${}^{\rho}\mathcal{D}^{\alpha,\beta}_{\tau_i^+}, {}^{\rho}\mathcal{J}^{1-\gamma}_{a^+}$ are the generalized Hilfer-type fractional derivative of order $\alpha \in (0, 1)$ and type $\beta \in [0, 1]$ and generalized fractional integral of order $1 - \gamma, (\gamma = \alpha + \beta - \alpha\beta)$, respectively, $\phi_1, \phi_2, \phi_3 \in \mathbb{R}, \phi_1 \neq 0, J_i := (\tau_i, t_{i+1}]; i = 0, \ldots, m, \tilde{J}_i := (t_i, \tau_i]; i = 1, \ldots, m, \quad a = t_0 = \tau_0 < t_1 \leq \tau_1 < t_2 \leq \tau_2 < \cdots \leq \tau_{m-1} < t_m \leq \tau_m < t_{m+1} = b < \infty, x(t_i^+) = \lim_{\epsilon \to 0^+} x(t_i + \epsilon)$ and $x(t_i^-) = \lim_{\epsilon \to 0^-} x(t_i + \epsilon)$ represent the right- and left-hand limits of $x(t)$ at $t = t_i$, $f : J_i \times \mathbb{R} \times \mathbb{R} \to \mathbb{R}$ is a given function, and $\psi_i : \tilde{J}_i \times \mathbb{R} \to \mathbb{R}; i = 1, \ldots, m$ are given continuous functions such that $\left({}^{\rho}\mathcal{J}^{1-\gamma}_{\tau_i^+} \psi_i \right)(t, x(t)) \big|_{t=\tau_i} = c_i \in \mathbb{R}$.

5.4.1   Existence Results

We can use the preliminary details, essential notations, definitions, and lemmas introduced in the two previous sections.

We consider the following linear fractional differential equation:

$$\left({}^{\rho}\mathcal{D}^{\alpha,\beta}_{\tau_i^+} x \right)(t) = v(t), \quad t \in J_i, i = 0, \ldots, m, \tag{5.43}$$

where $0 < \alpha < 1, 0 \leq \beta \leq 1, \rho > 0$, with the conditions

$$x(t) = \psi_i(t, x(\tau_i^-)); \ t \in \tilde{J}_i, \ i = 1, \ldots, m, \tag{5.44}$$

$$\phi_1 \left({}^\rho \mathcal{J}_{a^+}^{1-\gamma} x \right)(a^+) + \phi_2 \left({}^\rho \mathcal{J}_{\tau_m^+}^{1-\gamma} x \right)(b) = \phi_3, \tag{5.45}$$

where $\gamma = \alpha + \beta - \alpha\beta$, $\phi_1, \phi_2, \phi_3 \in \mathbb{R}$, $\phi_1 \neq 0$, and $c^* = max\{|c_i| : i = 1, \ldots, m\}$.

The following theorem shows that the problem (5.43)–(5.45) has a unique solution given by

$$x(t) = \begin{cases} \frac{1}{\Gamma(\gamma)} \left(\frac{t^\rho - a^\rho}{\rho} \right)^{\gamma-1} \left[\frac{\phi_3}{\phi_1} - \frac{c_m \phi_2}{\phi_1} - \frac{\phi_2}{\phi_1} \left({}^\rho \mathcal{J}_{\tau_m^+}^{1-\gamma+\alpha} v \right)(b) \right] \\ \qquad + \left({}^\rho \mathcal{J}_{a^+}^{\alpha} v \right)(t) \qquad if \ t \in J_0, \\ \\ \frac{c_i}{\Gamma(\gamma)} \left(\frac{t^\rho - \tau_i^\rho}{\rho} \right)^{\gamma-1} + \left({}^\rho \mathcal{J}_{\tau_i^+}^{\alpha} v \right)(t), \ t \in J_i, i = 1, \ldots, m, \\ \\ \psi_i(t, x(t)), \ t \in \tilde{J}_i, i = 1, \ldots, m. \end{cases} \tag{5.46}$$

Theorem 5.21 *Let* $\gamma = \alpha + \beta - \alpha\beta$, *where* $0 < \alpha < 1$ *and* $0 \le \beta \le 1$. *If* $v : J_i \to \mathbb{R}$; $i = 0, \ldots, m$, *is a function such that* $v(\cdot) \in C_{\gamma,\rho}(J_i)$, *then* $x \in PC_{\gamma,\rho}^\gamma(J)$ *satisfies the problem (5.43)–(5.45) if and only if it satisfies (5.46).*

Proof Assume x satisfies (5.43)–(5.45). If $t \in J_0$, then

$$\left({}^\rho \mathcal{D}_{a^+}^{\alpha,\beta} x \right)(t) = v(t).$$

Lemma 2.38 implies we have a solution that can be written as

$$x(t) = \frac{\left({}^\rho \mathcal{J}_{a^+}^{1-\gamma} x \right)(a)}{\Gamma(\gamma)} \left(\frac{t^\rho - a^\rho}{\rho} \right)^{\gamma-1} + \frac{1}{\Gamma(\alpha)} \int_a^t \left(\frac{t^\rho - \tau^\rho}{\rho} \right)^{\alpha-1} \tau^{\rho-1} v(\tau) d\tau.$$

If $t \in \tilde{J}_1$, then we have $x(t) = \psi_1(t, x(t))$.
If $t \in J_1$, then Lemma 2.38 implies

$$x(t) = \frac{\left({}^\rho \mathcal{J}_{\tau_1^+}^{1-\gamma} x \right)(\tau_1)}{\Gamma(\gamma)} \left(\frac{t^\rho - \tau_1^\rho}{\rho} \right)^{\gamma-1} + \frac{1}{\Gamma(\alpha)} \int_{\tau_1}^t \left(\frac{t^\rho - \tau^\rho}{\rho} \right)^{\alpha-1} \tau^{\rho-1} v(\tau) d\tau$$

$$= \frac{c_1}{\Gamma(\gamma)} \left(\frac{t^\rho - \tau_1^\rho}{\rho} \right)^{\gamma-1} + \left({}^\rho \mathcal{J}_{\tau_1^+}^{\alpha} v \right)(t).$$

If $t \in \tilde{J}_2$, then we have $x(t) = \psi_2(t, x(t))$.

If $t \in J_2$, then Lemma 2.38 implies

$$x(t) = \frac{\left(^{\rho}\mathcal{J}_{\tau_2^+}^{1-\gamma}x\right)(\tau_2)}{\Gamma(\gamma)}\left(\frac{t^{\rho}-\tau_2^{\rho}}{\rho}\right)^{\gamma-1} + \frac{1}{\Gamma(\alpha)}\int_{\tau_2}^{t}\left(\frac{t^{\rho}-\tau^{\rho}}{\rho}\right)^{\alpha-1}\tau^{\rho-1}v(\tau)d\tau$$

$$= \frac{c_2}{\Gamma(\gamma)}\left(\frac{t^{\rho}-\tau_2^{\rho}}{\rho}\right)^{\gamma-1} + \left(^{\rho}\mathcal{J}_{\tau_2^+}^{\alpha}v\right)(t).$$

Repeating the process in this way, the solution $x(t)$ for $t \in J$ can be written as

$$x(t) = \begin{cases} \dfrac{\left(^{\rho}\mathcal{J}_{a^+}^{1-\gamma}x\right)(a)}{\Gamma(\gamma)}\left(\dfrac{t^{\rho}-a^{\rho}}{\rho}\right)^{\gamma-1} + \left(^{\rho}\mathcal{J}_{a^+}^{\alpha}v\right)(t) & if\ t \in J_0, \\[3mm] \dfrac{c_i}{\Gamma(\gamma)}\left(\dfrac{t^{\rho}-\tau_i^{\rho}}{\rho}\right)^{\gamma-1} + \left(^{\rho}\mathcal{J}_{\tau_i^+}^{\alpha}v\right)(t) & if\ t \in J_i, i = 1,\ldots,m, \\[3mm] \psi_i(t,x(t)) & if\ t \in \tilde{J}_i, i = 1,\ldots,m. \end{cases}$$ (5.47)

Applying $^{\rho}\mathcal{J}_{\tau_m^+}^{1-\gamma}$ on both sides of (5.47), using Lemma 2.19, and taking $t = b$, we obtain

$$\left(^{\rho}\mathcal{J}_{\tau_m^+}^{1-\gamma}x\right)(b) = c_m + \left(^{\rho}\mathcal{J}_{\tau_m^+}^{1-\gamma+\alpha}v\right)(b).$$

Using the condition (5.45), we get

$$\left(^{\rho}\mathcal{J}_{a^+}^{1-\gamma}x\right)(a) = \frac{\phi_3}{\phi_1} - \frac{c_m\phi_2}{\phi_1} - \frac{\phi_2}{\phi_1}\left(^{\rho}\mathcal{J}_{\tau_m^+}^{1-\gamma+\alpha}v\right)(b).$$ (5.48)

Substituting (5.48) in (5.47), we get (5.46).

Reciprocally, for $t \in J_i$; $i = 0,\ldots,m$, applying $^{\rho}\mathcal{J}_{\tau_i^+}^{1-\gamma}$ on both sides of (5.46) and using Lemma 2.19 and Theorem 2.14, we get

$$\left(^{\rho}\mathcal{J}_{\tau_i^+}^{1-\gamma}x\right)(t) = \begin{cases} \dfrac{\phi_3}{\phi_1} - \dfrac{c_m\phi_2}{\phi_1} - \dfrac{\phi_2}{\phi_1}\left(^{\rho}\mathcal{J}_{\tau_m^+}^{1-\gamma+\alpha}v\right)(b) + \left(^{\rho}\mathcal{J}_{a^+}^{1-\gamma+\alpha}v\right)(t), t \in J_0, \\[3mm] c_i + \left(^{\rho}\mathcal{J}_{\tau_i^+}^{1-\gamma+\alpha}v\right)(t), t \in J_i, i = 1,\ldots,m. \end{cases}$$

(5.49)

Next, taking the limit $t \to a^+$ of (5.49) and using Lemma 2.24, with $1 - \gamma < 1 - \gamma + \alpha$, we obtain

$$\left({}^{\rho}\mathcal{J}_{a^{+}}^{1-\gamma} u \right)(a^{+}) = \frac{\phi_3}{\phi_1} - \frac{c_m \phi_2}{\phi_1} - \frac{\phi_2}{\phi_1} \left({}^{\rho}\mathcal{J}_{\tau_m^{+}}^{1-\gamma+\alpha} v \right)(b). \qquad (5.50)$$

Now taking $t = b$ in (5.49), we get

$$\left({}^{\rho}\mathcal{J}_{\tau_m^{+}}^{1-\gamma} u \right)(b) = c_m + \left({}^{\rho}\mathcal{J}_{\tau_m^{+}}^{1-\gamma+\alpha} v \right)(b). \qquad (5.51)$$

From (5.50) and (5.51), we obtain

$$\phi_1 \left({}^{\rho}\mathcal{J}_{a^{+}}^{1-\gamma} x \right)(a^{+}) + \phi_2 \left({}^{\rho}\mathcal{J}_{\tau_m^{+}}^{1-\gamma} x \right)(b) = \phi_3,$$

which shows that the boundary condition (5.45) is satisfied.

Next, for $t \in J_i$; $i = 0, \ldots, m$, apply operator ${}^{\rho}\mathcal{D}_{\tau_i^{+}}^{\gamma}$ on both sides of (5.46). Then, from Lemmas 2.19 and 2.33, we obtain

$$({}^{\rho}\mathcal{D}_{\tau_i^{+}}^{\gamma} x)(t) = \left({}^{\rho}\mathcal{D}_{\tau_i^{+}}^{\beta(1-\alpha)} v \right)(t). \qquad (5.52)$$

Since $x \in C_{\gamma,\rho}^{\gamma}(J_i)$ and by definition of $C_{\gamma,\rho}^{\gamma}(J_i)$, we have ${}^{\rho}\mathcal{D}_{\tau_i^{+}}^{\gamma} x \in C_{\gamma,\rho}(J_i)$, then (5.52) implies that

$$({}^{\rho}\mathcal{D}_{\tau_i^{+}}^{\gamma} x)(t) = \left(\delta_{\rho} \, {}^{\rho}\mathcal{J}_{\tau_i^{+}}^{1-\beta(1-\alpha)} v \right)(t) = \left({}^{\rho}\mathcal{D}_{\tau_i^{+}}^{\beta(1-\alpha)} v \right)(t) \in C_{\gamma,\rho}(J_i). \qquad (5.53)$$

As $v(\cdot) \in C_{\gamma,\rho}(J_i)$ and from Lemma 2.23, follows

$$\left({}^{\rho}\mathcal{J}_{\tau_i^{+}}^{1-\beta(1-\alpha)} v \right) \in C_{\gamma,\rho}(J_i), i = 0, \ldots, m. \qquad (5.54)$$

From (5.53), (5.54), and by the definition of the space $C_{\gamma,\rho}^{n}(J_i)$, we obtain

$$\left({}^{\rho}\mathcal{J}_{\tau_i^{+}}^{1-\beta(1-\alpha)} v \right) \in C_{\gamma,\rho}^{1}(J_i), i = 0, \ldots, m.$$

Applying operator ${}^{\rho}\mathcal{J}_{\tau_i^{+}}^{\beta(1-\alpha)}$ on both sides of (5.52) and using Lemmas 2.32 and 2.24 and Property 2.22, we have

$$\left({}^{\rho}\mathcal{D}_{\tau_i^{+}}^{\alpha,\beta} x \right)(t) = {}^{\rho}\mathcal{J}_{\tau_i^{+}}^{\beta(1-\alpha)} \left({}^{\rho}\mathcal{D}_{\tau_i^{+}}^{\gamma} x \right)(t)$$

$$= v(t) - \frac{\left({}^{\rho}\mathcal{J}_{\tau_i^{+}}^{1-\beta(1-\alpha)} v \right)(\tau_i)}{\Gamma(\beta(1-\alpha))} \left(\frac{t^{\rho} - \tau_i^{\rho}}{\rho} \right)^{\beta(1-\alpha)-1}$$

$$= v(t),$$

that is, (5.43) holds. Also, we can easily have

$$x(t) = \psi_i(t, x(t)); \ t \in \tilde{J}_i, \ i = 1, \ldots, m.$$

This completes the proof.   □

As a consequence of Theorem 5.21, we have the following result.

Lemma 5.22 *Let* $\gamma = \alpha + \beta - \alpha\beta$ *where* $0 < \alpha < 1$ *and* $0 \le \beta \le 1$, *and* $i = 0, \ldots, m$, *let* $f : J \times \mathbb{R} \times \mathbb{R} \to \mathbb{R}$ *be a function such that* $f(\cdot, x(\cdot), y(\cdot)) \in C_{\gamma,\rho}(J_i)$, *for any* $x, y \in PC_{\gamma,\rho}(J)$. *If* $x \in PC_{\gamma,\rho}^{\gamma}(J)$, *then* x *satisfies the problem* (5.40)–(5.42) *if and only if* x *is the fixed point of the operator* $\Im : PC_{\gamma,\rho}(J) \to PC_{\gamma,\rho}(J)$ *defined by*

$$\Im x(t) = \begin{cases} \dfrac{\overline{c}}{\Gamma(\gamma)} \left(\dfrac{t^\rho - a^\rho}{\rho} \right)^{\gamma-1} + \left({}^\rho \mathcal{J}_{a^+}^\alpha \varphi \right)(t), t \in J_0, \\[4mm] \dfrac{c_i}{\Gamma(\gamma)} \left(\dfrac{t^\rho - \tau_i^\rho}{\rho} \right)^{\gamma-1} + \left({}^\rho \mathcal{J}_{\tau_i^+}^\alpha \varphi \right)(t), \ t \in J_i, i = 1, \ldots, m, \\[4mm] \psi_i(t, x(t)), \ t \in \tilde{J}_i, i = 1, \ldots, m, \end{cases} \tag{5.55}$$

where φ *be a function satisfying the functional equation*

$$\varphi(t) = f(t, x(t), \varphi(t)),$$

and $\overline{c} = \dfrac{\phi_3}{\phi_1} - \dfrac{c_m \phi_2}{\phi_1} - \dfrac{\phi_2}{\phi_1} \left({}^\rho \mathcal{J}_{\tau_m^+}^{1-\gamma+\alpha} \varphi \right)(b)$. *Also, by Lemma 2.23,* $\Im u \in PC_{\gamma,\rho}(J)$.

We are now in a position to state and prove our existence result for the problem (5.40)–(5.42) based on Banach's fixed point theorem.

Theorem 5.23 *Assume that the following hypotheses hold.*

(5.23.1) *The function* $f : J_i \times \mathbb{R} \times \mathbb{R} \to \mathbb{R}$ *is continuous on* J_i; $i = 0, \ldots, m$, *and*

$$f(\cdot, x(\cdot), y(\cdot)) \in C_{\gamma,\rho}^{\beta(1-\alpha)}(J_i), i = 0, \ldots, m, \ \text{for any } x, y \in PC_{\gamma,\rho}(J).$$

(5.23.2) *There exist constants* $\eta_1 > 0$ *and* $0 < \eta_2 < 1$ *such that*

$$|f(t, x, y) - f(t, \bar{x}, \bar{y})| \le \eta_1 |x - \bar{x}| + \eta_2 |y - \bar{y}|$$

for any $x, y, \bar{x}, \bar{y} \in \mathbb{R}$ *and* $t \in J_i, i = 0, \ldots, m$.

(5.23.3) *The functions ψ_i are continuous and there exists a constant $K^* > 0$ such that*

$$|\psi_i(x) - \psi_i(\bar{x})| \leq K^*|x - \bar{x}|, \; x, \bar{x} \in \mathbb{R}, \; i = 1, \ldots, m.$$

If

$$\tilde{\ell} = \max\left\{K^*, \frac{\eta_1}{1 - \eta_2}\left(\frac{b^\rho - a^\rho}{\rho}\right)^\alpha \left[\frac{|\phi_2|}{|\phi_1|\Gamma(\alpha + 1)} + \frac{\Gamma(\gamma)}{\Gamma(\gamma + \alpha)}\right]\right\} < 1, \qquad (5.56)$$

then the problem (5.40)–(5.42) has a unique solution in $PC_{\gamma,\rho}(J)$.

Before starting the proof of Theorem 5.23, we are obliged to provide the following remark.

Remark 5.24 By hypothesis (5.23.2), we may have the following:

$$|f(t, x, y)| \leq |f(t, x, y) - f(t, 0, 0)| + |f(t, 0, 0)|$$
$$\leq \eta_1|x| + \eta_2|y| + f_0,$$

where $f_0 = \sup_{t \in [a,b]} |f(t, 0, 0)|$.

Proof The proof will be given in two steps.

Step 1: We show that the operator $\Im$ defined in (5.55) has a unique fixed point x^* in $PC_{\gamma,\rho}(J)$. Let $x, y \in PC_{\gamma,\rho}(J)$ and $t \in J$.
For $t \in J_0$ we have

$$|\Im x(t) - \Im y(t)| \leq \frac{|\phi_2|}{|\phi_1|\Gamma(\gamma)}\left(\frac{t^\rho - a^\rho}{\rho}\right)^{\gamma - 1}\left({}^\rho\mathcal{J}_{\tau_m^+}^{1-\gamma+\alpha}|\varphi(\tau) - \tilde{\varphi}(\tau)|\right) (b)$$
$$+ \left({}^\rho\mathcal{J}_{a^+}^{\alpha}|\varphi(\tau) - \tilde{\varphi}(\tau)|\right)(t),$$

and for $t \in J_i, i = 1, \ldots, m$, we have

$$|\Im x(t) - \Im y(t)| \leq \left({}^\rho\mathcal{J}_{\tau_i^+}^{\alpha}|\varphi(\tau) - \tilde{\varphi}(\tau)|\right)(t),$$

where $\varphi, \tilde{\varphi} \in C_{\gamma,\rho}(J_i); i = 0, \ldots, m$, such that

$$\varphi(t) = f(t, x(t), \varphi(t)),$$
$$\tilde{\varphi}(t) = f(t, y(t), \tilde{\varphi}(t)).$$

By (5.23.2), we have

$$|\varphi(t) - \tilde{\varphi}(t)| = |f(t, x(t), \varphi(t)) - f(t, y(t), \tilde{\varphi}(t))|$$
$$\leq \eta_1|x(t) - y(t)| + \eta_2|\varphi(t) - \tilde{\varphi}(t)|.$$

Then,

$$|\varphi(t) - \varphi(t)| \leq \frac{\eta_1}{1 - \eta_2} |x(t) - y(t)|.$$

Therefore, for each $t \in J_i, i = 1, \dots, m$,

$$|\Im x(t) - \Im y(t)| \leq \frac{\eta_1}{1 - \eta_2} \left({}^{\rho}\mathcal{J}_{\tau_i^+}^{\alpha} |x(\tau) - y(\tau)| \right)(t).$$

Thus

$$|\Im x(t) - \Im y(t)| \leq \left[\frac{\eta_1}{1 - \eta_2} \left({}^{\rho}\mathcal{J}_{\tau_i^+}^{\alpha} \left(\frac{\tau^{\rho} - \tau_i^{\rho}}{\rho} \right)^{\gamma - 1} \right)(t) \right] \|x - y\|_{PC_{\gamma,\rho}}.$$

By Lemma 2.19, we have

$$|\Im x(t) - \Im y(t)| \leq \left[\frac{\eta_1 \Gamma(\gamma)}{(1 - \eta_2)\Gamma(\gamma + \alpha)} \left(\frac{t^{\rho} - \tau_i^{\rho}}{\rho} \right)^{\alpha + \gamma - 1} \right] \|x - y\|_{PC_{\gamma,\rho}},$$

hence

$$\left| (\Im x(t) - \Im y(t)) \left(\frac{t^{\rho} - \tau_i^{\rho}}{\rho} \right)^{1 - \gamma} \right| \leq \left[\frac{\eta_1 \Gamma(\gamma) \left(\frac{t^{\rho} - \tau_i^{\rho}}{\rho} \right)^{\alpha}}{(1 - \eta_2)\Gamma(\gamma + \alpha)} \right] \|x - y\|_{PC_{\gamma,\rho}}$$

$$\leq \left[\frac{\eta_1 \Gamma(\gamma) \left(\frac{b^{\rho} - a^{\rho}}{\rho} \right)^{\alpha}}{(1 - \eta_2)\Gamma(\gamma + \alpha)} \right] \|x - y\|_{PC_{\gamma,\rho}}$$

$$\leq \tilde{\ell} \|x - y\|_{PC_{\gamma,\rho}}.$$

And for $t \in J_0$ we have

$$|\Im x(t) - \Im y(t)| \leq \frac{|\phi_2|}{|\phi_1|\Gamma(\gamma)} \left(\frac{t^{\rho} - a^{\rho}}{\rho} \right)^{\gamma - 1} \left({}^{\rho}\mathcal{J}_{\tau_m^+}^{1 - \gamma + \alpha} |\varphi(\tau) - \tilde{\varphi}(\tau)| \right)(b)$$

$$+ \left({}^{\rho}\mathcal{J}_{a^+}^{\alpha} |\varphi(\tau) - \tilde{\varphi}(\tau)| \right)(t)$$

$$\leq \left[\frac{|\phi_2|}{|\phi_1|\Gamma(\alpha + 1)} \left(\frac{b^{\rho} - \tau_m^{\rho}}{\rho} \right)^{\alpha + \gamma - 1} \right.$$

$$\left. + \frac{\Gamma(\gamma)}{\Gamma(\gamma + \alpha)} \left(\frac{t^{\rho} - a^{\rho}}{\rho} \right)^{\alpha + \gamma - 1} \right] \times \frac{\eta_1}{1 - \eta_2} \|x - y\|_{PC_{\gamma,\rho}},$$

hence

$$\left| (\Im x(t) - \Im y(t)) \left(\frac{t^\rho - a^\rho}{\rho} \right)^{1-\gamma} \right| \leq \frac{\eta_1 (b^\rho - a^\rho)^\alpha}{(1 - \eta_2)\rho^\alpha} \left[\frac{|\phi_2|}{|\phi_1|\Gamma(\alpha + 1)} + \frac{\Gamma(\gamma)}{\Gamma(\gamma + \alpha)} \right]$$

$$\times \|x - y\|_{PC_{\gamma,\rho}}$$

$$\leq \tilde{\ell}\|x - y\|_{PC_{\gamma,\rho}}.$$

For $t \in \tilde{J}_i, i = 1, \ldots, m$, we have

$$|\Im x(t) - \Im y(t)| \leq |(\psi_i(t, x(t)) - \psi_i(t, y(t)))|$$

$$\leq K^* \|x - y\|_{PC_{\gamma,\rho}}$$

$$\leq \tilde{\ell}\|x - y\|_{PC_{\gamma,\rho}}.$$

Then, for each $t \in J$, we have

$$\|\Im x - \Im y\|_{PC_{\gamma,\rho}} \leq \tilde{\ell}\|u - w\|_{PC_{\gamma,\rho}}.$$

By (5.56), the operator $\Im$ is a contraction. Hence, by Theorem 2.45, $\Im$ has a unique fixed point $x^* \in PC_{\gamma,\rho}(J)$.

Step 2: We prove that the fixed point $x^* \in PC_{\gamma,\rho}(J)$ is actually in $PC^\gamma_{\gamma,\rho}(J)$.
Since x^* is the unique fixed point of operator $\Im$ in $PC_{\gamma,\rho}(J)$, then for each $t \in J$, we have

$$\Im x^*(t) = \begin{cases} \dfrac{\bar{c}}{\Gamma(\gamma)} \left(\dfrac{t^\rho - a^\rho}{\rho} \right)^{\gamma-1} + \left({}^\rho\mathcal{J}^\alpha_{a^+}\varphi \right)(t) & \textit{if } t \in J_0, \\[3mm] \dfrac{c_i}{\Gamma(\gamma)} \left(\dfrac{t^\rho - \tau_i^\rho}{\rho} \right)^{\gamma-1} + \left({}^\rho\mathcal{J}^\alpha_{\tau_i^+}\varphi \right)(t) & \textit{if } t \in J_i, i = 1, \ldots, m, \\[3mm] \psi_i(t, x^*(t)) & \textit{if } t \in \tilde{J}_i, i = 1, \ldots, m, \end{cases}$$

where $\varphi \in C_{\gamma,\rho}(J_i); i = 0, \ldots, m$, such that

$$\varphi(t) = f(t, x^*(t), \varphi(t)).$$

For $t \in J_i; i = 0, \ldots, m$, applying ${}^\rho\mathcal{D}^\gamma_{\tau_i^+}$ to both sides and by Lemmas 2.19 and 2.33, we have

$${}^\rho\mathcal{D}^\gamma_{\tau_i^+}x^*(t) = \left({}^\rho\mathcal{D}^\gamma_{\tau_i^+} {}^\rho\mathcal{J}^\alpha_{\tau_i^+} f(\tau, x^*(\tau), \varphi(\tau)) \right)(t)$$

$$= \left({}^\rho\mathcal{D}^{\beta(1-\alpha)}_{\tau_i^+} f(\tau, x^*(\tau), \varphi(\tau)) \right)(t).$$

Since $\gamma \geq \alpha$, by (5.23.1), the right-hand side is in $C_{\gamma,\rho}(J_i)$ and thus ${}^\rho\mathcal{D}^\gamma_{\tau_i^+}x^* \in C_{\gamma,\rho}(J_i)$.
And since $\psi_i \in C(\tilde{J}_i, \mathbb{R}); i = 1, \ldots, m$, then $x^* \in PC^\gamma_{\gamma,\rho}(J)$. As a consequence of Steps

1 and 2 together with Theorem 5.23, we can conclude that the problem (5.40)–(5.42) has a unique solution in $PC_{\gamma,\rho}(J)$. $\qquad\square$

Our second result is based on Krasnoselskii's fixed point theorem.

Theorem 5.25 *Assume that (5.23.1), (5.23.2), and the following condition hold:*

(5.25.1) *The functions ψ_i are continuous and there exist constants $0 < \Phi_1 < 1$, $\Phi_2 > 0$ such that*

$$|\psi_i(x)| \le \Phi_1|x| + \Phi_2 \text{ for each } x \in \mathbb{R}, i = 1, \ldots, m.$$

If

$$\frac{|\phi_2|\eta_1}{|\phi_1|\Gamma(\alpha+1)(1-\eta_2)} \left(\frac{b^\rho - a^\rho}{\rho}\right)^\alpha < 1, \tag{5.57}$$

then the problem (5.40)–(5.42) has at least one solution in $PC_{\gamma,\rho}(J)$.

Proof Consider the set

$$B_\omega = \{x \in PC_{\gamma,\rho}(J) : \|x\|_{PC_{\gamma,\rho}} \le \omega\},$$

where $\omega \ge r_1 + r_2$, with

$$r_1 := \max \left\{ \frac{c^*}{\Gamma(\gamma)}, \frac{|\phi_3 - c_m\phi_2|}{\Gamma(\gamma)|\phi_1|} + \frac{A|\phi_2|}{\Gamma(\alpha+1)|\phi_1|} \left(\frac{b^\rho - a^\rho}{\rho}\right)^\alpha \right\},$$

$$r_2 := \max \left\{ \Phi_1 r + \Phi_2, A \left(\frac{\Gamma(\gamma)}{\Gamma(\gamma+\alpha)}\right) \left(\frac{b^\rho - a^\rho}{\rho}\right)^\alpha \right\}.$$

We define the operators N_1 and N_2 on B_ω by

$$N_1 x(t) = \begin{cases} \dfrac{1}{\Gamma(\gamma)} \left(\dfrac{t^\rho - a^\rho}{\rho}\right)^{\gamma-1} \left[\dfrac{\phi_3}{\phi_1} - \dfrac{c_m\phi_2}{\phi_1} - \dfrac{\phi_2}{\phi_1} \left({}^\rho J_{\tau_m^+}^{1-\gamma+\alpha}\varphi\right)(b)\right], & t \in J_0, \\[3mm] \dfrac{c_i}{\Gamma(\gamma)} \left(\dfrac{t^\rho - \tau_i^\rho}{\rho}\right)^{\gamma-1}, & t \in J_i, i = 1, \ldots, m, \\[3mm] 0, & t \in \tilde{J}_i, i = 1, \ldots, m \end{cases} \tag{5.58}$$

and

$$N_2x(t) = \begin{cases} \left(^\rho \mathcal{J}^\alpha_{\tau_i^+} \varphi \right)(t) & if \ t \in J_i, 0 = 1, \ldots, m, \\[2ex] \psi_i(t, x(t)) & if \ t \in \tilde{J}_i, i = 1, \ldots, m \end{cases} \tag{5.59}$$

where $i = 0, \ldots, m$ and $\varphi : J_i \rightarrow \mathbb{R}$ be a function satisfying the functional equation

$$\varphi(t) = f(t, x(t), \varphi(t)).$$

Then the fractional integral equation (5.55) can be written as operator equation

$$\Im x(t) = N_1 x(t) + N_2 x(t), \quad x \in PC_{\gamma,\rho}(J).$$

We shall use Krasnoselskii's fixed point theorem to prove in several steps that the operator $\Im$ defined in (5.55) has a fixed point.

Step 1: We prove that $N_1 x + N_2 y \in B_\omega$ for any $x, y \in B_\omega$.
By Remark (5.24) and from (5.55), we have for each $t \in J_i, i = 0, \ldots, m$,

$$\left| \left(\frac{t^\rho - \tau_i^\rho}{\rho} \right)^{1-\gamma} \varphi(t) \right| = \left| \left(\frac{t^\rho - \tau_i^\rho}{\rho} \right)^{1-\gamma} f(t, x(t), \varphi(t)) \right|$$

$$\leq \left(\frac{t^\rho - \tau_i^\rho}{\rho} \right)^{1-\gamma} (\eta_1 |x(t)| + \eta_2 |\varphi(t)| + f_0),$$

which implies that

$$\left| \left(\frac{t^\rho - \tau_i^\rho}{\rho} \right)^{1-\gamma} \varphi(t) \right| \leq \eta_1 \left(\frac{b^\rho - a^\rho}{\rho} \right)^{1-\gamma} \omega + \eta_2 \left| \left(\frac{t^\rho - \tau_i^\rho}{\rho} \right)^{1-\gamma} \varphi(t) \right|$$

$$+ f_0 \left(\frac{b^\rho - a^\rho}{\rho} \right)^{1-\gamma}.$$

Then

$$\max_{i=0,\ldots,m} \left\{ \sup_{t \in J_i} \left| \left(\frac{t^\rho - \tau_i^\rho}{\rho} \right)^{1-\gamma} \varphi(t) \right| \right\} \leq \frac{(\eta_1 \omega + f_0) \left(\frac{b^\rho - a^\rho}{\rho} \right)^{1-\gamma}}{1 - \eta_2} := A.$$

Thus, for $t \in J_0$, by (5.58) and Lemma 2.19,

$$\left| \left(\frac{t^\rho - a^\rho}{\rho} \right)^{1-\gamma} (N_1 x)(t) \right| \leq \frac{|\phi_3 - c_m \phi_2|}{\Gamma(\gamma)|\phi_1|} + \frac{|\phi_2|}{\Gamma(\gamma)|\phi_1|} \left({}^\rho \mathcal{J}_{\tau_m^+}^{1-\gamma+\alpha} |\varphi(\tau)| \right) (b)$$

$$\leq \frac{|\phi_3 - c_m \phi_2|}{\Gamma(\gamma)|\phi_1|} + \frac{A|\phi_2|}{\Gamma(\alpha+1)|\phi_1|} \left(\frac{b^\rho - a^\rho}{\rho} \right)^\alpha,$$

and for $t \in J_i, i = 1, \ldots, m$, we have

$$\left| \left(\frac{t^\rho - \tau_i^\rho}{\rho} \right)^{1-\gamma} (N_1 x)(t) \right| \leq \frac{|c_i|}{\Gamma(\gamma)} \leq \frac{c^*}{\Gamma(\gamma)},$$

then for each $t \in J$ we get

$$\|N_1 x\|_{PC_{\gamma,\rho}} \leq \max \left\{ \frac{c^*}{\Gamma(\gamma)}, \frac{|\phi_3 - c_m \phi_2|}{\Gamma(\gamma)|\phi_1|} + \frac{A|\phi_2|}{\Gamma(\alpha+1)|\phi_1|} \left(\frac{b^\rho - a^\rho}{\rho} \right)^\alpha \right\}. \quad (5.60)$$

For $t \in J_i, i = 0, \ldots, m$, by (5.59) and Lemma 2.19, we have

$$\left| \left(\frac{t^\rho - \tau_i^\rho}{\rho} \right)^{1-\gamma} (N_2 y)(t) \right| \leq \left(\frac{t^\rho - \tau_i^\rho}{\rho} \right)^{1-\gamma} \left({}^\rho \mathcal{J}_{\tau_i^+}^\alpha |\varphi(\tau)| \right) (t)$$

$$\leq A \left(\frac{\Gamma(\gamma)}{\Gamma(\gamma+\alpha)} \right) \left(\frac{b^\rho - a^\rho}{\rho} \right)^\alpha,$$

and for each $t \in \tilde{J}_i, i = 1, \ldots, m$, we have

$$|(N_2 y)(t)| \leq |\psi_i(t, y(t))|$$
$$\leq \Phi_1 r + \Phi_2,$$

then for each $t \in J$, we get

$$\|N_2 y\|_{PC_{\gamma,\rho}} \leq \max \left\{ \Phi_1 r + \Phi_2, A \left(\frac{\Gamma(\gamma)}{\Gamma(\gamma+\alpha)} \right) \left(\frac{b^\rho - a^\rho}{\rho} \right)^\alpha \right\}. \quad (5.61)$$

From (5.60) and (5.61), for each $t \in J$, we have

$$\|N_1 x + N_2 y\|_{PC_{\gamma,\rho}} \leq \|N_1 x\|_{PC_{\gamma,\rho}} + \|N_2 y\|_{PC_{\gamma,\rho}}$$
$$\leq r_1 + r_2$$
$$\leq \omega,$$

which infers that $N_1 x + N_2 y \in B_\omega$.

Step 2: N_1 is a contraction.
Let $x, y \in PC_{\gamma,\rho}(J)$ and $t \in J$.
By (5.23.2), we have

$$|\varphi(t) - \tilde{\varphi}(t)| = |f(t, x(t), \varphi(t)) - f(t, y(t), \tilde{\varphi}(t))|$$
$$\leq \eta_1 |x(t) - y(t)| + \eta_2 |\varphi(t) - \tilde{\varphi}(t)|.$$

where $\varphi, \tilde{\varphi} \in C_{\gamma,\rho}(J_i); i = 0, \ldots, m$, such that

$$\varphi(t) = f(t, x(t), \varphi(t)),$$
$$\tilde{\varphi}(t) = f(t, y(t), \tilde{\varphi}(t)).$$

Then,

$$|\varphi(t) - \varphi(t)| \leq \frac{\eta_1}{1 - \eta_2} |x(t) - y(t)|.$$

Therefore, for $t \in J_0$, we have

$$|N_1 x(t) - N_1 y(t)| \leq \frac{|\phi_2|}{|\phi_1| \Gamma(\gamma)} \left(\frac{t^\rho - a^\rho}{\rho}\right)^{\gamma-1} \left({}^\rho \mathcal{J}_{\tau_m^+}^{1-\gamma+\alpha} |\varphi(\tau) - \tilde{\varphi}(\tau)|\right) (b)$$

$$\leq \frac{\eta_1}{1 - \eta_2} \left[\frac{|\phi_2|}{|\phi_1| \Gamma(\alpha + 1)} \left(\frac{b^\rho - \tau_m^\rho}{\rho}\right)^{\alpha+\gamma-1}\right] \|x - y\|_{PC_{\gamma,\rho}}.$$

Hence,

$$\left|(N_1 x(t) - N_1 y(t)) \left(\frac{t^\rho - a^\rho}{\rho}\right)^{1-\gamma}\right| \leq \frac{|\phi_2| \eta_1 \left(\frac{b^\rho - a^\rho}{\rho}\right)^\alpha}{|\phi_1| \Gamma(\alpha + 1)(1 - \eta_2)} \|x - y\|_{PC_{\gamma,\rho}}.$$

Then, for each $t \in J$, we have

$$\|N_1 x - N_1 y\|_{PC_{\gamma,\rho}} \leq \frac{|\phi_2| \eta_1}{|\phi_1| \Gamma(\alpha + 1)(1 - \eta_2)} \left(\frac{b^\rho - a^\rho}{\rho}\right)^\alpha \|x - y\|_{PC_{\gamma,\rho}}.$$

Then by (5.57), the operator N_1 is a contraction.

Step 3: N_2 is continuous and compact. Let $\{x_n\}$ be a sequence such that $x_n \to x$ in $PC_{\gamma,\rho}(J)$. Then for each $t \in J_i, i = 0, \ldots, m$, we have

$$\left|(N_2 x_n)(t) - (N_2 x)(t)) \left(\frac{t^\rho - \tau_i^\rho}{\rho}\right)^{1-\gamma}\right| \leq \left(\frac{t^\rho - \tau_i^\rho}{\rho}\right)^{1-\gamma} \left({}^\rho \mathcal{J}_{\tau_i^+}^\alpha |\varphi_n(\tau) - \varphi(\tau)|\right)(t),$$

where $h_n, h \in C(J_i, \mathbb{R})$, such that

$$\varphi_n(t) = f(t, x_n(t), \varphi_n(t)),$$
$$\varphi(t) = f(t, x(t), \varphi(t)).$$

For each $t \in \tilde{J}_i, i = 1, \ldots, m$, we have

$$|(N_2 x_n)(t) - (N_2 x)(t)| \leq |(\psi_i(t, x_n(t)) - \psi_i(t, x(t)))|.$$

Since $x_n \to x$, then we get $\varphi_n(t) \to \varphi(t)$ as $n \to \infty$ for each $t \in J_i; i = 0, \dots, m$. By Lebesgue's dominated convergence Theorem and since ψ_i are continuous, we have

$$\|N_2 x_n - N_2 x\|_{PC_{\gamma,\rho}} \to 0 \text{ as } n \to \infty.$$

Then N_2 is continuous. Next, we prove that N_2 is uniformly bounded on B_ω. Let any $y \in B_\omega$. We have from step 1 that for each $t \in J$

$$\|N_2 y\|_{PC_{\gamma,\rho}} \le \max \left\{ \Phi_1 r + \Phi_2, A \left(\frac{\Gamma(\gamma)}{\Gamma(\gamma + \alpha)} \right) \left(\frac{b^\rho - a^\rho}{\rho} \right)^\alpha \right\}.$$

This proves that the operator N_2 is uniformly bounded on B_ω. To prove the compactness of N_2, we take $y \in B_\omega$ and $a < \varepsilon_1 < \varepsilon_2 \le b$. Then for $\varepsilon_1, \varepsilon_2 \in J_i; i = 0, \dots, m$,

$$\left| \left(\frac{\varepsilon_1^\rho - \tau_i^\rho}{\rho} \right)^{1-\gamma} (N_2 y)(\varepsilon_1) - \left(\frac{\varepsilon_2^\rho - \tau_i^\rho}{\rho} \right)^{1-\gamma} (N_2 y)(\varepsilon_2) \right|$$

$$\le \left| \left(\frac{\varepsilon_1^\rho - \tau_i^\rho}{\rho} \right)^{1-\gamma} \left({}^\rho \mathcal{J}_{\tau_i^+}^\alpha \varphi(\tau) \right)(\varepsilon_1) - \left(\frac{\varepsilon_2^\rho - \tau_i^\rho}{\rho} \right)^{1-\gamma} \left({}^\rho \mathcal{J}_{\tau_i^+}^\alpha \varphi(\tau) \right)(\varepsilon_2) \right|$$

$$\le \left(\frac{\varepsilon_2^\rho - \tau_i^\rho}{\rho} \right)^{1-\gamma} \left({}^\rho \mathcal{J}_{\varepsilon_1^+}^\alpha |\varphi(\tau)| \right)(\varepsilon_2) + \frac{1}{\Gamma(\alpha)} \int_{\tau_i}^{\varepsilon_1} \left| \tau^{\rho-1} H(\tau) \varphi(\tau) \right| d\tau,$$

where

$$H(\tau) = \left[\left(\frac{\varepsilon_1^\rho - \tau_i^\rho}{\rho} \right)^{1-\gamma} \left(\frac{\varepsilon_1^\rho - \tau^\rho}{\rho} \right)^{\alpha-1} - \left(\frac{\varepsilon_2^\rho - \tau_i^\rho}{\rho} \right)^{1-\gamma} \left(\frac{\varepsilon_2^\rho - \tau^\rho}{\rho} \right)^{\alpha-1} \right].$$

Then by Lemma 2.19, we have

$$\left| \left(\frac{\varepsilon_1^\rho - \tau_i^\rho}{\rho} \right)^{1-\gamma} (N_2 y)(\varepsilon_1) - \left(\frac{\varepsilon_2^\rho - \tau_i^\rho}{\rho} \right)^{1-\gamma} (N_2 y)(\varepsilon_2) \right|$$

$$\le \frac{A\Gamma(\gamma)}{\Gamma(\alpha + \gamma)} \left(\frac{\varepsilon_2^\rho - \tau_i^\rho}{\rho} \right)^{1-\gamma} \left(\frac{\varepsilon_2^\rho - \varepsilon_1^\rho}{\rho} \right)^{\alpha+\gamma-1}$$

$$+ A \int_{\tau_i}^{\varepsilon_1} \left| H(\tau) \frac{\tau^{\rho-1}}{\Gamma(\alpha)} \right| \left(\frac{\tau^\rho - \tau_i^\rho}{\rho} \right)^{\gamma-1} d\tau,$$

note that

$$\left| \left(\frac{\varepsilon_1^\rho - \tau_i^\rho}{\rho} \right)^{1-\gamma} (N_2 y)(\varepsilon_1) - \left(\frac{\varepsilon_2^\rho - \tau_i^\rho}{\rho} \right)^{1-\gamma} (N_2 y)(\varepsilon_2) \right| \to 0 \text{ as } \varepsilon_1 \to \varepsilon_2.$$

And for $\varepsilon_1, \varepsilon_2 \in \tilde{J}_i; i = 1, \ldots, m$,

$$|(N_2 y)(\varepsilon_1) - (N_2 y)(\varepsilon_2)| \leq |\psi_i(\varepsilon_1, y(\varepsilon_1)) - \psi_i(\varepsilon_2, y(\varepsilon_2))|,$$

note since ψ_i are continuous that

$$|(N_2 y)(\varepsilon_1) - (N_2 y)(\varepsilon_2)| \to 0 \quad \text{as} \quad \varepsilon_1 \to \varepsilon_2.$$

This proves that $N_2 B_\omega$ is equicontinuous on J. Therefore, $N_2 B_\omega$ is relatively compact. By $PC_{\gamma,\rho}$-type Arzela-Ascoli Theorem, N_2 is compact. As a consequence of Theorem 2.50, we deduce that $\mathfrak{I}$ has at least a fixed point $x^* \in PC_{\gamma,\rho}(J)$, and by the same way of the proof of Theorem 5.23, we can easily show that $x^* \in PC_{\gamma,\rho}^{\gamma}(J)$. Using Lemma 5.22, we conclude that the problem (5.40)–(5.42) has at least one solution in the space $PC_{\gamma,\rho}(J)$. $\square$

5.4.2 Ulam-Hyers-Rassias Stability

Now, we consider the Ulam stability for problem (5.40)–(5.42). Let $x \in PC_{\gamma,\rho}(J), \theta > 0$, $\mu > 0$, and $\chi : J \longrightarrow [0, \infty)$ be a continuous function. We consider the following inequality:

$$\begin{cases} \left| \left({}^{\rho}\mathcal{D}_{\tau_i^+}^{\alpha,\beta} x \right)(t) - f\left(t, x(t), \left({}^{\rho}\mathcal{D}_{\tau_i^+}^{\alpha,\beta} x\right)(t)\right) \right| \leq \theta, t \in J_i, i = 0, \ldots, m, \\[2mm] |x(t) - \psi_i(t, x(t))| \leq \theta, t \in \tilde{J}_i, i = 1, \ldots, m, \end{cases} \tag{5.62}$$

$$\begin{cases} \left| \left({}^{\rho}\mathcal{D}_{\tau_i^+}^{\alpha,\beta} x \right)(t) - f\left(t, x(t), \left({}^{\rho}\mathcal{D}_{\tau_i^+}^{\alpha,\beta} x\right)(t)\right) \right| \leq \chi(t), t \in J_i, i = 0, \ldots, m, \\[2mm] |x(t) - \psi_i(t, x(t))| \leq \mu, t \in \tilde{J}_i, i = 1, \ldots, m, \end{cases} \tag{5.63}$$

and

$$\begin{cases} \left| \left({}^{\rho}\mathcal{D}_{\tau_i^+}^{\alpha,\beta} x \right)(t) - f\left(t, x(t), \left({}^{\rho}\mathcal{D}_{\tau_i^+}^{\alpha,\beta} x\right)(t)\right) \right| \leq \theta\chi(t), t \in J_i, i = 0, \ldots, m, \\[2mm] |x(t) - \psi_i(t, x(t))| \leq \theta\mu, t \in \tilde{J}_i, i = 1, \ldots, m. \end{cases} \tag{5.64}$$

Definition 5.26 ([156, 158]) Problem (5.40)–(5.42) is Ulam-Hyers (U-H) stable if there exists a real number $a_f > 0$ such that for each $\theta > 0$ and for each solution $x \in PC_{\gamma,\rho}(J)$ of inequality (5.62) there exists a solution $y \in PC_{\gamma,\rho}(J)$ of (5.40)–(5.42) with

$$|x(t) - y(t)| \leq \theta a_f, \quad t \in J.$$

Definition 5.27 ([156, 158]) Problem (5.40)–(5.42) is generalized Ulam-Hyers (G.U-H) stable if there exists $K_f : C([0, \infty), [0, \infty))$ with $K_f(0) = 0$ such that for each $\theta > 0$ and for each solution $x \in PC_{\gamma,\rho}(J)$ of inequality (5.62) there exists a solution $y \in PC_{\gamma,\rho}(J)$ of (5.40)–(5.42) with

$$|x(t) - y(t)| \le K_f(\theta), \qquad t \in J.$$

Definition 5.28 ([156, 158]) Problem (5.40)–(5.42) is Ulam-Hyers-Rassias (U-H-R) stable with respect to (χ, μ) if there exists a real number $a_{f,\chi} > 0$ such that for each $\theta > 0$ and for each solution $x \in PC_{\gamma,\rho}(J)$ of inequality (5.64) there exists a solution $y \in PC_{\gamma,\rho}(J)$ of (5.40)–(5.42) with

$$|x(t) - y(t)| \le \theta a_{f,\chi}(\chi(t) + \mu), \qquad t \in J.$$

Definition 5.29 ([156, 158]) Problem (5.40)–(5.42) is generalized Ulam-Hyers-Rassias (G.U-H-R) stable with respect to (χ, μ) if there exists a real number $a_{f,\chi} > 0$ such that for each solution $x \in PC_{\gamma,\rho}(J)$ of inequality (5.63) there exists a solution $y \in PC_{\gamma,\rho}(J)$ of (5.40)–(5.42) with

$$|x(t) - y(t)| \le a_{f,\chi}(\chi(t) + \mu), \qquad t \in J.$$

Remark 5.30 It is clear that

1. Definition 5.26 $\implies$ Definition 5.27.
2. Definition 5.28 $\implies$ Definition 5.29.
3. Definition 5.28 for $\chi(.) = \mu = 1 \implies$ Definition 5.26.

Remark 5.31 ([156, 158]) A function $x \in PC_{\gamma,\rho}(J)$ is a solution of inequality (5.64) if and only if there exist $\upsilon \in PC_{\gamma,\rho}(J)$ and a sequence $\upsilon_i, i = 0, \ldots, m$ such that

1. $|\upsilon(t)| \le \theta \chi(t), t \in J_i, i = 0, \ldots, m$; and $|\upsilon_i| \le \theta \mu, t \in \tilde{J}_i, i = 1, \ldots, m$,
2. $\left(^\rho D_{\tau_i^+}^{\alpha,\beta} x\right)(t) = f\left(t, x(t), \left(^\rho D_{\tau_i^+}^{\alpha,\beta} x\right)(t)\right) + \upsilon(t), t \in J_i, i = 0, \ldots, m$,
3. $x(t) = \psi_i(t, x(t)) + \upsilon_i, t \in \tilde{J}_i, i = 1, \ldots, m$.

Theorem 5.32 *Assume that in addition to (5.23.1)–(5.23.3) and (5.56), the following hypothesis holds:*

(5.32.1) *There exist a nondecreasing function* $\chi : J \longrightarrow [0, \infty)$ *and* $\kappa_\chi > 0$ *such that for each* $t \in J_i; i = 0, \ldots, m$, *we have*

$$(^\rho J_{\tau_i^+}^\alpha \chi)(t) \le \kappa_\chi \chi(t).$$

Then the problem (5.40)–(5.42) is U-H-R stable with respect to (χ, μ).

Proof Let $x \in PC_{\gamma,\rho}(J)$ be a solution of inequality (5.64), and let us assume that y is the unique solution of the problem

$$
\begin{cases}
\left({}^\rho D^{\alpha,\beta}_{\tau^+_i} y \right)(t) = f\left(t, y(t), \left({}^\rho D^{\alpha,\beta}_{\tau^+_i} y \right)(t)\right); \ t \in J_i, \ i = 0, \ldots, m, \\
y(t) = \psi_i(t, y(t)); \ t \in \tilde{J}_i, \ i = 1, \ldots, m, \\
\phi_1 \left({}^\rho J^{1-\gamma}_{a^+} y \right)(a^+) + \phi_2 \left({}^\rho J^{1-\gamma}_{m^+} y \right)(b) = \phi_3, \\
\left({}^\rho J^{1-\gamma}_{\tau^+_i} y \right)(\tau_i) = \left({}^\rho J^{1-\gamma}_{\tau^+_i} x \right)(\tau_i), \ i = 0, \ldots, m.
\end{cases}
$$

By Lemma 5.22, we obtain for each $t \in J$

$$
y(t) = \begin{cases}
\dfrac{\bar{c}}{\Gamma(\gamma)} \left(\dfrac{t^\rho - a^\rho}{\rho} \right)^{\gamma-1} + \left({}^\rho J^{\alpha}_{a^+}\varphi \right)(t), \ t \in J_0, \\[3mm]
\dfrac{\left({}^\rho J^{1-\gamma}_{\tau^+_i} y \right)(\tau_i)}{\Gamma(\gamma)} \left(\dfrac{t^\rho - \tau^\rho_i}{\rho} \right)^{\gamma-1} + \left({}^\rho J^{\alpha}_{\tau^+_i}\varphi \right)(t), \ t \in J_i, i = 1, \ldots, m, \\[3mm]
\psi_i(t, y(t)), \ t \in \tilde{J}_i, i = 1, \ldots, m,
\end{cases}
$$

where $\varphi \in C_{\gamma,\rho}(J_i); i = 0, \ldots, m,$ be a function satisfying the functional equation

$$\varphi(t) = f(t, y(t), \varphi(t))$$

and $\bar{c} = \dfrac{\phi_3}{\phi_1} - \dfrac{\phi_2}{\phi_1} \left({}^\rho J^{1-\gamma}_{\tau^+_m} y \right)(\tau_m) - \dfrac{\phi_2}{\phi_1} \left({}^\rho J^{1-\gamma+\alpha}_{\tau^+_m}\varphi \right)(b).$

Since x is a solution of the inequality (5.64), by Remark 5.31, we have

$$
\begin{cases}
\left({}^\rho D^{\alpha,\beta}_{\tau^+_i} x \right)(t) = f\left(t, x(t), \left({}^\rho D^{\alpha,\beta}_{\tau^+_i} x \right)(t)\right) + \upsilon(t), t \in J_i, i = 0, \ldots, m; \\
x(t) = \psi_i(t, x(t)) + \upsilon_i, t \in \tilde{J}_i, i = 1, \ldots, m.
\end{cases} \tag{5.65}
$$

Clearly, the solution of (5.65) is given by

$$
x(t) = \begin{cases}
\dfrac{\left({}^\rho J^{1-\gamma}_{\tau^+_i} x \right)(\tau_i)}{\Gamma(\gamma)} \left(\dfrac{t^\rho - \tau^\rho_i}{\rho} \right)^{\gamma-1} \\[3mm]
\quad + \left({}^\rho J^{\alpha}_{\tau^+_i}(\tilde{\varphi} + \upsilon) \right)(t) \ if \ t \in J_i, i = 1, \ldots, m, \\[3mm]
\psi_i(t, x(t)) + \upsilon_i \quad if \ t \in \tilde{J}_i, i = 1, \ldots, m,
\end{cases}
$$

where $\tilde{\varphi} : J_i \to \mathbb{R}, i = 0, \ldots, m$, be a function satisfying the functional equation

$$\tilde{\varphi}(t) = f(t, x(t), \tilde{\varphi}(t)).$$

Hence, for each $t \in J_i, i = 0, \ldots, m$, we have

$$|x(t) - y(t)| \leq \left({}^\rho J_{\tau_i^+}^\alpha |\tilde{\varphi}(\tau) - \varphi(\tau)| \right)(t) + \left({}^\rho J_{\tau_i^+}^\alpha |\upsilon(\tau)| \right)$$

$$\leq \theta \kappa_\chi \chi(t) + \frac{\eta_1}{(1 - \eta_2)} \int_{\tau_i}^t \tau^{\rho-1} \left(\frac{t^\rho - \tau^\rho}{\rho} \right)^{\alpha-1} \frac{|x(\tau) - y(\tau)|}{\Gamma(\alpha)} d\tau.$$

We apply Lemma 2.40 to obtain

$$|x(t) - y(t)| \leq \theta \kappa_\chi \chi(t) + \int_{\tau_i}^t \sum_{k=1}^\infty \frac{\left(\frac{\eta_1}{1-\eta_2} \right)^k}{\Gamma(k\alpha)} \tau^{\rho-1} \left(\frac{t^\rho - \tau^\rho}{\rho} \right)^{k\alpha-1} (\theta \kappa_\chi \chi(\tau)) d\tau$$

$$\leq \theta \kappa_\chi \chi(t) E_\alpha \left[\frac{\eta_1}{1-\eta_2} \left(\frac{t^\rho - \tau_i^\rho}{\rho} \right)^\alpha \right]$$

$$\leq \theta \kappa_\chi \chi(t) E_\alpha \left[\frac{\eta_1}{1-\eta_2} \left(\frac{b^\rho - a^\rho}{\rho} \right)^\alpha \right].$$

And for each $t \in \tilde{J}_i, i = 1, \ldots, m$, we have

$$|x(t) - y(t)| \leq |\psi_i(t, x(t)) - \psi_i(t, y(t))| + |\upsilon_i|$$
$$\leq K^* |x(t) - y(t)| + \theta \mu,$$

then by 5.56, we have

$$|x(t) - y(t)| \leq \frac{\theta \mu}{1 - K^*}.$$

Then for each $t \in J$, we have

$$|x(t) - y(t)| \leq a_\chi \theta(\mu + \chi(t)),$$

where

$$a_\chi = \frac{1}{1 - K^*} + \kappa_\chi E_\alpha \left[\frac{\eta_1}{1-\eta_2} \left(\frac{b^\rho - a^\rho}{\rho} \right)^\alpha \right].$$

Hence, the problem (5.40)–(5.42) is U-H-R stable with respect to (χ, τ). $\square$

Remark 5.33 If the conditions (5.23.1)–(5.23.3), (5.32.1), and (5.56) are satisfied, then by Theorem 5.32 and Remark 5.30, it is clear that problem (5.40)–(5.42) is U-H-R stable and G.U-H-R stable. And if $\chi(.) = \mu = 1$, then problem (5.40)–(5.42) is also G.U-H stable and U-H stable.

Remark 5.34 Our results for the boundary value problem (5.40)–(5.42) apply in the following cases:

- Initial value problems: $\phi_1 = 1, \phi_2 = 0$.
- Anti-periodic problems: $\phi_1 = 1, \phi_2 = 1, \phi_3 = 0$.
- Periodic problems: $\phi_1 = 1, \phi_2 = -1, \phi_3 = 0$.

5.4.3 An Example

Example 5.35 Consider the following impulsive periodic problem of generalized Hilfer fractional differential equation:

$$\left(\tfrac{1}{2} D_{\tau_i^+}^{\frac{1}{2},0} x\right)(t) = \frac{|\cos(t)|e^{-2t} + |\sin(t)|}{122 e^{t+2}(1 + |x(t)| + |\tfrac{1}{2} D_{\tau_i^+}^{\frac{1}{2},0} x(t)|)}, \quad \text{for each } t \in J_0 \cup J_1, \qquad (5.66)$$

$$x(t) = \frac{|x(t)|}{5e^t + 3|x(t)|}, \quad \text{for each } t \in \tilde{J}_1, \qquad (5.67)$$

$$\left(\tfrac{1}{2} J_{1^+}^{\frac{1}{2}} x\right)(1^+) = \left(\tfrac{1}{2} J_{3^+}^{\frac{1}{2}} x\right)(\pi), \qquad (5.68)$$

where $J_0 = (1, e], J_1 = (3, \pi], \tilde{J}_1 = (e, 3], s_0 = 1, t_1 = e,$ and $s_1 = 3$.
Set

$$f(t, u, w) = \frac{|\cos(t)|e^{-2t} + |\sin(t)|}{122 e^{t+2}(1 + |x| + |y|)}, \quad t \in J_0 \cup J_1, \ x, y \in \mathbb{R}.$$

We have

$$C_{\gamma,\rho}^{\beta(1-\alpha)}((1, e]) = C_{\frac{1}{2},\frac{1}{2}}^0((1, e])$$

$$= \left\{ u : (1, e] \to \mathbb{R} : \sqrt{2}\left(\sqrt{t} - 1\right)^{\frac{1}{2}} u \in C([1, e], \mathbb{R}) \right\},$$

and

$$C_{\gamma,\rho}^{\beta(1-\alpha)}((3, \pi]) = C_{\frac{1}{2},\frac{1}{2}}^0((3, \pi])$$

$$= \left\{ u : (3, \pi] \to \mathbb{R} : \sqrt{2}\left(\sqrt{t} - \sqrt{3}\right)^{\frac{1}{2}} u \in C([3, \pi], \mathbb{R}) \right\},$$

with

$$\gamma = \alpha = \frac{1}{2}, \ \rho = \frac{1}{2}, \ \beta = 0, \ and \ i \in \{0, 1\}.$$

Clearly, the continuous function $f \in C_{\frac{1}{2},\frac{1}{2}}^{0}((1, e]) \cap C_{\frac{1}{2},\frac{1}{2}}^{0}((3, \pi])$. Hence, the condition (5.23.1) is satisfied.

For each $x, \bar{x}, y, \bar{y} \in \mathbb{R}$ and $t \in J_0 \cup J_1$, we have

$$|f(t, x, y) - f(t, \bar{x}, \bar{y})| \le \frac{|cos(t)|e^{-2t} + |sin(t)|}{122e^{t+2}}(|x - \bar{x}| + |y - \bar{y}|)$$

$$\le \frac{1 + e^2}{122e^5} (|x - \bar{x}| + |y - \bar{y}|).$$

Hence condition (5.23.2) is satisfied with $\eta_1 = \eta_2 = \dfrac{1 + e^2}{122e^5}$.

And let

$$\psi(x) = \frac{x}{5e^t + 3x}, u \in [0, \infty).$$

Let $x, y \in [0, \infty)$. Then we have

$$|\psi(x) - \psi(y)| = |\frac{x}{5e^t + 3x} - \frac{y}{5e^t + 3y}| = \frac{5e^t |x - y|}{(5e^t + 3x)(5e^t + 3y)} \le \frac{1}{5e}|x - y|,$$

and so the condition (5.23.3) is satisfied with $K^* = \dfrac{1}{5e}$.

Also, the condition (5.56) of Theorem 5.23 is satisfied for

$$\tilde{\ell} = \max \left\{ K^*, \frac{\eta_1}{1 - \eta_2} \left(\frac{b^\rho - a^\rho}{\rho}\right)^\alpha \left[\frac{|\phi_2|}{|\phi_1|\Gamma(\alpha + 1)} + \frac{\Gamma(\gamma)}{\Gamma(\gamma + \alpha)}\right] \right\}$$

$$= \max \left\{ \frac{1}{5e}, \frac{\sqrt{2}(1 + e^2)}{122e^5 - e^2 - 1} (\sqrt{\pi} - 1)^{\frac{1}{2}} \left[\frac{1}{\Gamma(\frac{3}{2})} + \sqrt{\pi}\right] \right\}$$

$$\approx \max \{0.0735758882, 0.00167130655\}$$

$$= 0.00167130655 < 1.$$

Then the problem (5.66)–(5.68) has a unique solution in $PC_{\frac{1}{2},\frac{1}{2}}([1, \pi])$.
Hypothesis (5.32.1) is satisfied with $\mu = 1$ and

$$\chi(t) = \begin{cases} \dfrac{1}{\sqrt{2(\sqrt{t} - \sqrt{\tau_i})}}, & \text{if } t \in J_0 \cup J_1, \\ \\ \pi, & \text{if } t \in \tilde{J}_1, \end{cases}$$

and $\kappa_\chi = \sqrt{2\pi}(\sqrt{e} - 1)^{\frac{1}{2}}$. Indeed, for each $t \in J_0 \cup J_1$, we get

$$(^{\frac{1}{2}}\mathcal{J}_{1+}^{\frac{1}{2}}\chi)(t) \le \frac{\sqrt{2\pi}(\sqrt{\pi} - \sqrt{3})^{\frac{1}{2}}}{\sqrt{2(\sqrt{t} - 1)}},$$

and

$$({}^{\frac{1}{2}}\mathcal{J}_{3+}^{\frac{1}{2}}\chi)(t) \leq \frac{\sqrt{2\pi}(\sqrt{e}-1)^{\frac{1}{2}}}{\sqrt{2(\sqrt{t}-\sqrt{3})}}.$$

Consequently, Theorem 5.32 implies that the problem (5.66)–(5.68) is U-H-R stable.

5.5   Notes and Remarks

The results of Chapter 5 are taken from the papers of Salim et al. [123, 128, 130]. We refer the reader to the monographs [7, 24, 27, 43, 61, 68, 70, 82, 97, 151], and the papers [1–6, 8, 15, 51, 52, 54–56], for more information on the concepts studied in this chapter.

References

1. S. Abbas, W. Albarakati, M. Benchohra and J.J. Nieto, Existence and stability results for partial implicit fractional differential equations with not instantaneous impulses. *Novi Sad J. Math.* **47** (2017), 157–171.
2. S. Abbas, M. Benchohra, Uniqueness and Ulam stabilities results for partial fractional differential equations with not instantaneous impulses. *Appl. Math. Comput.* **257** (2015), 190–198.
3. S. Abbas and M. Benchohra, Stability results for fractional differential equations with state-dependent delay and not instantaneous impulses. *Math. Slovaca.* **67** (2017), 875–894.
4. S. Abbas, M. Benchohra, Uniqueness and Ulam stabilities results for partial fractional differential equations with not instantaneous impulses. *Appl. Math. Comput.* **257** (2015), 190–198.
5. S. Abbas, M. Benchohra, A. Alsaedi and Y. Zhou, Some stability concepts for abstract fractional differential equations with not instantaneous impulses. *Fixed Point Theory.* **18** (2017), 3–16.
6. S. Abbas, M. Benchohra, M.A. Darwish, New stability results for partial fractional differential inclusions with not instantaneous impulses. *Frac. Calc. Appl. Anal.* **18** (2015), 172–191.
7. S. Abbas, M. Benchohra, J. R. Graef and J. Henderson, *Implicit Differential and Integral Equations: Existence and stability*, Walter de Gruyter, London, 2018.
8. S. Abbas, M. Benchohra, J. R. Graef, J. E. Lazreg, Implicit Hadamard fractional differential equations with impulses under weak topologies. *Dyn. Contin. Discrete Impuls. Syst. Ser. A Math. Anal.* **26** (2019), 89–112.
9. S. Abbas, M. Benchohra, J. Henderson and J. E. Lazreg, Measure of noncompactness and impulsive Hadamard fractional implicit differential equations in Banach spaces. *Math. Eng. Science Aerospace.* **8** (2017), 1–19.
10. S. Abbas, M. Benchohra, J. E. Lazreg, A Alsaedi and Y. Zhou, Existence and Ulam stability for fractional differential equations of Hilfer-Hadamard type. *Adv. Difference Equ.* **2017** (2017), 180.
11. S. Abbas, M. Benchohra, J. E. Lazreg and G. N'Guérékata, Hilfer and Hadamard functional random fractional differential inclusions. *Cubo* **19** (2017), 17–38.
12. S. Abbas, M. Benchohra, J. E. Lazreg and Y. Zhou, A survey on Hadamard and Hilfer fractional differential equations: analysis and stability. *Chaos Solitons Fractals.* **102** (2017), 47–71.

M. Benchohra et al., *Advanced Topics in Fractional Differential Equations*,
Synthesis Lectures on Mathematics & Statistics,
https://doi.org/10.1007/978-3-031-26928-8

13. S. Abbas, M. Benchohra and G. M. N'Guérékata, *Advanced Fractional Differential and Integral Equations*, Nova Science Publishers, New York, 2014.

14. S. Abbas, M. Benchohra and G. M. N'Guérékata, *Topics in Fractional Differential Equations*, Springer-Verlag, New York, 2012.

15. S. Abbas, M. Benchohra and J. J. Trujillo, Upper and lower solutions method for partial fractional differential inclusions with not instantaneous impulses. *Prog. Frac. Differ. Appl.* **1** (2015), 11–22.

16. S. Abbas, M. Benchohra and S. Sivasundaram, Coupled Pettis-Hadamard fractional differential systems with retarded and advanced arguments , *J. Math. Stat.* **14** (2018), 56–63.

17. T. Abdeljawad, R.P. Agarwal, E. Karapınar and P.S. Kumari, Solutions of the Nonlinear Integral Equation and Fractional Differential Equation Using the Technique of a Fixed Point with a Numerical Experiment in Extended b-Metric Space. *Symmetry.* **11** (2019), 686.

18. S.A. Abd-Salam, A.M.A. El-Sayed, On the stability of a fractional-order differential equation with nonlocal initial condition. *Electron. J. Qual. Theory Differ. Equ.* **29** (2008), 1–8.

19. R.S. Adiguzel, U. Aksoy, E. Karapınar and I.M. Erhan, On the solution of a boundary value problem associated with a fractional differential equation. *Math. Methods Appl. Sci.* **2020** (2020), 1– 12.

20. H. Afshari, S. Kalantari and D. Baleanu , Solution of fractional differential equations via $\alpha - \psi$-Geraghty type mappings. *Adv. Difference Equ.* **2018** (2018), 347.

21. H. Afshari, S. Kalantari and E. Karapınar, Solution of fractional differential equations via coupled FP. *Electron. J. Differential Equations.* **2015** (2015), 1–12.

22. R. Agarwal, Certain fractional q-integrals and q-derivatives. *Proc. Camb. Philos. Soc.* **66** (1969), 365–370.

23. P. Agarwal, D. Baleanu, Y. Chen, S. Momani, and J. A. T. Machado, *Fractional Calculus: ICFDA 2018, Amman, Jordan, July 16–18*, Springer, Singapore, 2019.

24. R.P. Agarwal, S. Hristova, D. O'Regan, *Non-Instantaneous Impulses in Differential Equations*, Springer, New York, 2017.

25. B. Ahmad, A. Alsaedi, S.K. Ntouyas and J. Tariboon, *Hadamard-type Fractional Differential Equations, Inclusions and Inequalities*. Springer, Cham, 2017.

26. B. Ahmad, J.R. Graef, Coupled systems of nonlinear fractional differential equations with nonlocal boundary conditions. *Panamer. Math. J.* **19** (2009), 29–39.

27. B. Ahmad, J. Henderson, and R. Luca, *Boundary value problems for fractional differential equations and systems*, World Scientific, USA, 2021.

28. B. Ahmad and S.K Ntouyas, Fractional differential inclusions with fractional separated boundary conditions. *Fract. Calc. Appl. Anal.* **15** (2012), 362–382.

29. B. Ahmad, S.K Ntouyas, Initial value problems for hybrid Hadamard fractional differential equations. *Electron. J. Differ. Equ.* **2014**, (2014), 161.

30. M.A. Almalahi, O. Bazighifan, S.K. Panchal, S.S. Askar and G.I. Oros, Analytical study of two nonlinear coupled hybrid systems involving generalized Hilfer fractional operators, *Fractal Fract.* **5** (2021), 22pp.

31. A. Almalahi and K. Panchal, Existence results of ψ-Hilfer integro-differential equations with fractional order in Banach space, *Ann. Univ. Paedagog. Crac. Stud. Math*, **19** (2020), 171–192.

32. R. Almeida, A Gronwall inequality for a general Caputo fractional operator. *Math. Inequal. Appl.* **20** (2017), 1089–1105.

33. R. Almeida, A.B. Malinowska and T. Odzijewicz, Fractional differential equations with dependence on the Caputo–Katugampola derivative. *J. Comput. Nonlinear Dynam.* **11** (2016), 1–11.

34. B. Alqahtani, H. Aydi, E. Karapınar and V. Rakocevic, A Solution for Volterra Fractional Integral Equations by Hybrid Contractions. *Mathematics.* **7** (2019), 694.

35. J. C. Alvàrez, Measure of noncompactness and fixed points of nonexpansive condensing mappings in locally convex spaces. *Rev. Real. Acad. Cienc. Exact. Fis. Natur. Madrid.* **79** (1985), 53–66.

36. A. Ali, K. Shah, R. A. Khan, Existence of solution to a coupled system of hybrid fractional differential equations, *Bull. Math. Anal. Appl.* **9** (2017), 9–18.

37. G. A. Anastassiou, *Generalized Fractional Calculus: New Advancements and Applications*, Springer International Publishing, Switzerland, 2021.

38. J. M. Ayerbee Toledano, T. Dominguez Benavides and G. Lopez Acedo, *Measures of Noncompactness in Metric Fixed Point Theory*, Operator Theory: Advances and Applications, Berlin, 1997.

39. L. Bai, J. J. Nieto and J. M. Uzal, On a delayed epidemic model with non-instantaneous impulses. *Commun. Pure Appl. Anal.* **19** (2020), no. 4, 1915–1930.

40. Z. Baitiche, M. Benbachir, K. Guerbati, Solvability of two-point fractional boundary value problems at resonance. *Malaya J. Mat.* **8** (2020), no. 2, 464–468.

41. Z. Baitiche, K. Guerbati, M. Benchohra, Y. Zhou, Boundary value problems for hybrid Caputo fractional differential equations. *Mathematics.* **2019** (2019), 282.

42. D. Baleanu, K. Diethelm, E. Scalas, and J.J. Trujillo, *Fractional Calculus Models and Numerical Methods*, World Scientific Publishing, New York, 2012.

43. D. Baleanu, Z.B. Güvenç, and J.A.T. Machado *New Trends in Nanotechnology and Fractional Calculus Applications*, Springer, New York, 2010.

44. S. Banach, Sur les opérations dans les ensembles abstraits et leur application aux équations intégrales, *Fund. Math.* **3** (1922), 133–181.

45. J. Banas and K. Goebel, *Measures of noncompactness in Banach spaces*. Marcel Dekker, New York, 1980.

46. H. Belbali, M. Benbachir, Stability for coupled systems on networks with Caputo-Hadamard fractional derivative. *J. Math. Model.* **9** (2021), no. 1, 107–118.

47. M. Benchohra, F. Bouazzaoui, E. Karapinar and A. Salim, Controllability of second order functional random differential equations with delay. *Mathematics.* **10** (2022), 16pp. https://doi.org/10.3390/math10071120

48. M. Benchohra, S. Bouriah and M. A. Darwish, Nonlinear boundary value problem for implicit differential equations of fractional order in Banach spaces. *Fixed Point Theory.* **18** (2017), No. 2, 457–470.

49. M. Benchohra, S. Bouriah and John R. Graef, Boundary value problems for nonlinear implicit Caputo-Hadamard-type fractional differential equations with impulses. *Mediterr. J. Math.* **14** (2017), 206.

50. M. Benchohra, J. Henderson, S. K. Ntouyas, *Impulsive Differential Equations and Inclusions*, Hindawi Publishing Corporation, New York, 2006.

51. M. Benchohra and J. E. Lazreg, Existence and Ulam stability for Nonlinear implicit fractional differential equations with Hadamard derivative. *Stud. Univ. Babes-Bolyai Math.* **62** (2017), 27–38.

52. M. Benchohra and J. E. Lazreg, On stability for nonlinear implicit fractional differential equations. *Matematiche (Catania).* **70** (2015), 49–61.

53. M. Benchohra and J. E. Lazreg, Existence results for nonlinear implicit fractional differential equations with impulse. *Commun. Appl. Anal.* **19** (2015), 413–426.

54. N. Benkhettou, K. Aissani, A. Salim, M. Benchohra and C. Tunc, Controllability of fractional integro-differential equations with infinite delay and non-instantaneous impulses, *Appl. Anal. Optim.* **6** (2022), 79–94.

55. N. Benkhettou, A. Salim, K. Aissani, M. Benchohra and E. Karapinar, Non-instantaneous impulsive fractional integro-differential equations with state-dependent delay, *Sahand Commun. Math. Anal.* **19** (2022), 93–109. https://doi.org/10.22130/scma.2022.542200.1014

56. A. Bensalem, A. Salim, M. Benchohra and G. N'Guérékata, Functional integro-differential equations with state-dependent delay and non-instantaneous impulsions: existence and qualitative results, *Fractal Fract.* **6** (2022), 1–27. https://doi.org/10.3390/fractalfract6100615

57. S. Bouriah, A. Salim and M. Benchohra, On nonlinear implicit neutral generalized Hilfer fractional differential equations with terminal conditions and delay, *Topol. Algebra Appl.* **10** (2022), 77–93. https://doi.org/10.1515/taa-2022-0115

58. A. Boutiara, M.S. Abdo, M. Benbachir, Existence results for ψ-Caputo fractional neutral functional integro-differential equations with finite delay. *Turkish J. Math.* **44** (2020), no. 6, 2380–2401.

59. A. Boutiara, K. Guerbati, M. Benbachir, Caputo-Hadamard fractional differential equation with three-point boundary conditions in Banach spaces. *AIMS Math.* **5** (2020), no. 1, 259–272.

60. L. Byszewski and V. Lakshmikantham, Theorem about the existence and uniqueness of a solution of a nonlocal abstract Cauchy problem in a Banach space. *Appl.Anal.* **40**(1991), 11–19.

61. E. Capelas de Oliveira, *Solved Exercises in Fractional Calculus*, Springer International Publishing, Switzerland, 2019.

62. K. Cao and Y. Chen, *Fractional Order Crowd Dynamics: Cyber-Human System Modeling and Control*, Berlin; Boston, De Gruyter, 2018.

63. Y. M. Chu, M. U. Awan, S. Talib, M. A. Noor and K. I. Noor, Generalizations of Hermite-Hadamard like inequalities involving χ_k-Hilfer fractional integrals, *Adv. Difference Equ.* **2020** (2020), 594.

64. C. Derbazi, H. Hammouche, M. Benchohra, et *al.* Fractional hybrid differential equations with three-point boundary hybrid conditions. *Adv. Difference Equ.* **2019**, (2019), 125.

65. C. Derbazi, H. Hammouche, A. Salim and M. Benchohra, Measure of noncompactness and fractional hybrid differential equations with hybrid conditions. *Differ. Equ. Appl.* **14** (2022), 145–161. http://dx.doi.org/10.7153/dea-2022-14-09

66. R. Diaz and C. Teruel, q, k-Generalized gamma and beta functions, *J. Nonlinear Math. Phys* **12** (2005), 118–134.

67. K. Diethelm, *The Analysis of Fractional Differential Equations*, Lecture Notes in Mathematics, 2010.

68. H. Dutta, A. O. Akdemir, and A. Atangana, *Fractional order analysis: theory, methods and applications*, Hoboken, NJ: Wiley, 2020.

69. P. Egbunonu, M. Guay, Identification of switched linear systems using subspace and integer programming techniques.*Nonlinear Anal. Hybrid Syst.* **1** (2007), 577–592.

70. M. Francesco, *Fractional Calculus: Theory and Applications*, MDPI, 2018.

71. K. Goebel, *Concise course on Fixed Point Theorems*. Yokohama Publishers, Japan, 2002.

72. A. Granas and J. Dugundji, *Fixed Point Theory*, Springer-Verlag, New York, 2003.

73. J. R. Graef, J. Henderson and A. Ouahab, *Impulsive differential inclusions. A Fixed Point Approch*, De Gruyter, Berlin/Boston, 2013.

74. K. Guida, K. Hilal and L. Ibnelazyz, Existence of mild solutions for a class of impulsive Hilfer fractional coupled systems, *Adv. Math. Phys.* **2020** (2020), 12pp.

75. D.J. Guo, V. Lakshmikantham, X. Liu, *Nonlinear Integral Equations in Abstract Spaces*. Kluwer Academic Publishers, Dordrecht, 1996.

76. S. Harikrishnan, R. W. Ibrahim and K. Kanagarajan. Fractional Ulam-stability of fractional impulsive differential equation involving Hilfer-Katugampola fractional differential operator. *Univ. J. Math. Appl.* **1** (2018), 106–112.

77. A. Heris, A. Salim, M. Benchohra and E. Karapinar, Fractional partial random differential equations with infinite delay. *Results in Physics*. (2022). https://doi.org/10.1016/j.rinp.2022.105557

78. E. Hernández, K. A. G. Azevedo and M. C. Gadotti, Existence and uniqueness of solution for abstract differential equations with state-dependent delayed impulses. *J. Fixed Point Theory Appl.* **21** (2019), no. 1, Paper No. 36, 17 pp.

79. K. Hilal, A. Kajouni, Boundary value problems for hybrid differential equations with fractional order. *Adv. Difference Equ.* **2015**, (2015), 183.

80. D. H. Hyers. On the stability of the linear functional equation. *Proc. Natl. Acad. Sci.* **27** (1941), 222–224.

81. B. Jin, *Fractional Differential Equations: An Approach via Fractional Derivatives*, Springer International Publishing, Switzerland, 2021.

82. G. Karniadakis, *Handbook of Fractional Calculus with Applications. Volume 3: Numerical Methods*, Berlin, Boston, De Gruyter, 2019.

83. E. Karapınar, T.Abdeljawad and F. Jarad, Applying new fixed point theorems on fractional and ordinary differential equations, *Adv. Difference Equ.* **2019** (2019), 421.

84. E. Karapınar, A. Fulga, M. Rashid, L. Shahid and H. Aydi, Large Contractions on Quasi-Metric Spaces with an Application to Nonlinear Fractional Differential-Equations. *Mathematics.* **7** (2019), 444.

85. A.A. Kilbas, H. M. Srivastava, and Juan J. Trujillo, *Theory and Applications of Fractional Differential Equations*. North-Holland Mathematics Studies, Amsterdam, 2006.

86. A. Kochubei, Y. Luchko, *Handbook of Fractional Calculus with Applications. Volume 1: Basic Theory*, Berlin, Boston, De Gruyter, 2019.

87. F. Kong and J. J. Nieto, Control of bounded solutions for first-order singular differential equations with impulses. *IMA J. Math. Control Inform.* **37** (2020), no. 3, 877–893.

88. S. Krim, A. Salim, S. Abbas and M. Benchohra, On implicit impulsive conformable fractional differential equations with infinite delay in b-metric spaces. *Rend. Circ. Mat. Palermo (2).* (2022), 1–14. https://doi.org/10.1007/s12215-022-00818-8

89. K. D. Kucche and A. D. Mali, Initial time difference quasilinearization method for fractional differential equations involving generalized Hilfer fractional derivative. *Comput. Appl. Math.* **39** (2020), no. 31, 1–33.

90. N. Laledj, A. Salim, J. E. Lazreg, S. Abbas, B. Ahmad and M. Benchohra, On implicit fractional q-difference equations: Analysis and stability. *Math. Methods Appl. Sci.* **45** (2022), no. 17, 10775–10797. https://doi.org/10.1002/mma.8417

91. J. E. Lazreg, M. Benchohra and A. Salim, Existence and Ulam stability of k-generalized ψ-Hilfer fractional problem. *J. Innov. Appl. Math. Comput. Sci.* **2** (2022), 01–13.

92. L. Lin, Y. Liu and D. Zhao, Study on implicit-type fractional coupled system with integral boundary conditions, *Math.* **9** (2021), 15pp.

93. J. Liouville, Second mémoire sur le développement des fonctions ou parties de fonctions en séries dont divers termes sont assujettis á satisfaire à une même équation différentielle du second ordre contenant un paramétre variable, *J. Math. Pure et Appi.* **2** (1837), 16–35.

94. K. Liu, J. Wang and D. O'Regan, Ulam-Hyers-Mittag-Leffler stability for ψ-Hilfer fractional-order delay differential equations, *Adv. Difference Equ*, **2019** (2019), 50.

95. A. J. Luo and V. Afraimovich, *Long-range Interactions, Stochasticity and Fractional Dynamics*, Springer, New York, Dordrecht, Heidelberg, London, 2010.

96. A. D. Mali and K. D. Kucche, Nonlocal boundary value problem for generalized Hilfer implicit fractional differential equations. *Math. Meth. Appl. Sci.* **43** (2020), no.15, 8608–8631.

97. R. Martínez-Guerra, F. Meléndez-Vázquez, and I. Trejo-Zuniga, *Fault-tolerant Control and Diagnosis for Integer and Fractional-order Systems: Fundamentals of Fractional Calculus and Differential Algebra with Real-Time Applications*, Springer International Publishing, 2021.
98. M. M. Meerschaert and A. Sikorskii, *Stochastic Models for Fractional Calculus*, De Gruyter, 2019.
99. K. S. Miller and B. Ross, *An Introduction to the Fractional Calculus and Differential Equations*, John Wiley, New York, 1993.
100. C. Milici, G. Draganescu, and J. A. T. Machado, *Introduction to Fractional Differential Equations*, Springer International Publishing, 2019.
101. H. Mönch, Boundary value problems for nonlinear ordinary differential equations of second order in Banach spaces, *Nonlinear Anal.* **4** (1980), 985–999.
102. S. Mubeen and G. M. Habibullah, k-Fractional Integrals and Application, *Int. J. Contemp. Math. Sciences*,**7** (2012), 89–94.
103. J. E. Nápoles Valdés, Generalized fractional Hilfer integral and derivative, *Contr. Math.* **2** (2020), 55–60.
104. S. Naz and M. N. Naeem, On the Generalization of k-Fractional Hilfer-Katugampola Derivative with Cauchy Problem, *Turk. J. Math.* **45** (2021), 110–124.
105. G M. N'Guérékata, A Cauchy problem for some fractional abstract differential equation with non local conditions. *Nonlinear Anal.* **70** (2009), 1873–1876.
106. D.S. Oliveira, E. Capelas de Oliveira, Hilfer–Katugampola fractional derivatives. *Comput. Appl. Math.* **37** (2018), 3672–3690.
107. K.B. Oldham and J. Spanier, *The Fractional Calculus : theory and application of differentiation and integration to arbitrary order*, Academic Press, New York, London, 1974.
108. M. D. Ortigueira, *Fractional Calculus for Scientists and Engineers*. Lecture Notes in Electrical Engineering, 84. Springer, Dordrecht, 2011.
109. D. O'Regan, Fixed point theory for weakly sequentially continuous mapping. *Math. Comput. Model.* **27** (1998), 1–14.
110. E. Picard, Mémoire sur la théorie des équations aux derivées partielles et la méthode des approximations successives, *J. Math. Pures et Appl.* **6** (1890), 145–210.
111. I. Podlubny, *Fractional Differential Equations*, Academic Press, San Diego, 1999.
112. H. Poincaré, Sur les courbes definies par les équations différentielles, *J. Math.* **2** (1886), 54–65.
113. S. Rashid, M. Aslam Noor, K. Inayat Noor, Y. M. Chu, Ostrowski type inequalities in the sense of generalized $\mathcal{K}$-fractional integral operator for exponentially convex functions,*AIMS Math.* **5** (2020), 2629–2645.
114. TM. Rassias. On the stability of the linear mappings in Banach spaces. *Proc. Amer. Math. Soc.* **72** (1978), 297–300.
115. Th. M. Rassias and J. Brzdek, *Functional Equations in Mathematical Analysis*, Springer 86, New York Dordrecht Heidelberg London 2012.
116. Th. M. Rassias, On the stability of the linear mapping in Banach spaces. *Proc. Amer. Math. Soc.* **72** (1978), 297–300.
117. J. M. Rassias, *Functional Equations, Difference Inequalities and Ulam Stability Notions (F.U.N.)*, Nova Science Publishers, Inc. New York, 2010.
118. B. Ross, *Fractional Calculus and Its Applications*, Proceedings of the International Conference, New Haven, Springer-Verlag, New York, 1974.
119. IA. Rus, Ulam stability of ordinary differential equations. *Stud. Univ. Babes-Bolyai. Math.* **4** (2009), 125–133.
120. A. Salim, S. Abbas, M. Benchohra and E. Karapinar, A Filippov's theorem and topological structure of solution sets for fractional q-difference inclusions. *Dynam. Systems Appl.* **31** (2022), 17–34. https://doi.org/10.46719/dsa202231.01.02

121. A. Salim, S. Abbas, M. Benchohra and E. Karapinar, Global stability results for Volterra-Hadamard random partial fractional integral equations. *Rend. Circ. Mat. Palermo (2).* (2022), 1–13. https://doi.org/10.1007/s12215-022-00770-7

122. A. Salim, B. Ahmad, M. Benchohra and J. E. Lazreg, Boundary value problem for hybrid generalized Hilfer fractional differential equations, *Differ. Equ. Appl.* **14** (2022), 379–391. http://dx.doi.org/10.7153/dea-2022-14-27

123. A. Salim, M. Benchohra, J. R. Graef and J. E. Lazreg, Boundary value problem for fractional generalized Hilfer-type fractional derivative with non-instantaneous impulses. *Fractal Fract.* **5** (2021), 1–21. https://dx.doi.org/10.3390/fractalfract5010001

124. A. Salim, M. Benchohra, J. R. Graef and J. E. Lazreg, Initial value problem for hybrid ψ-Hilfer fractional implicit differential equations. *J. Fixed Point Theory Appl.* **24** (2022), 14 pp. https://doi.org/10.1007/s11784-021-00920-x

125. A. Salim, M. Benchohra, E. Karapinar and J. E. Lazreg, Existence and Ulam stability for impulsive generalized Hilfer-type fractional differential equations. *Adv. Differ. Equ.* **2020** (2020), 21 pp. https://doi.org/10.1186/s13662-020-03063-4

126. A. Salim, M. Benchohra and J. E. Lazreg, Boundary value problem for differential equations with Hilfer-Katugampola fractional derivative in Banach spaces. (Submitted).

127. A. Salim, M. Benchohra and J. E. Lazreg, k-Generalized ψ-Hilfer differential equations in Banach spaces. (Submitted).

128. A. Salim, M. Benchohra and J. E. Lazreg, Nonlinear implicit generalized Hilfer type fractional differential equations with non-instantaneous impulses. (Submitted).

129. A. Salim, M. Benchohra and J. E. Lazreg, Nonlocal k-generalized ψ-Hilfer impulsive initial value problem with retarded and advanced arguments, *Appl. Anal. Optim.* **6** (2022), 21–47.

130. A. Salim, M. Benchohra, J. E. Lazreg and J. Henderson, Nonlinear implicit generalized Hilfer-type fractional differential equations with non-instantaneous impulses in Banach spaces. *ATNAA.* **4** (2020), 332–348. https://doi.org/10.31197/atnaa.825294

131. A. Salim, M. Benchohra, J. E. Lazreg and J. Henderson, On k-generalized ψ-Hilfer boundary value problems with retardation and anticipation. *ATNAA.* **6** (2022), 173–190. https://doi.org/10.31197/atnaa.973992

132. A. Salim, M. Benchohra, J. E. Lazreg and E. Karapınar, On k-generalized ψ-Hilfer impulsive boundary value problem with retarded and advanced arguments. *J. Math. Ext.* **15** (2021), 1–39. https://doi.org/10.30495/JME.SI.2021.2187

133. A. Salim, M. Benchohra, J. E. Lazreg and G. N'Guérékata, Boundary value problem for nonlinear implicit generalized Hilfer-type fractional differential equations with impulses. *Abstr. Appl. Anal.* **2021** (2021), 17pp. https://doi.org/10.1155/2021/5592010

134. A. Salim, M. Benchohra, J. E. Lazreg and G. N'Guérékata, Existence and k-Mittag-Leffler-Ulam-Hyers stability results of k-generalized ψ-Hilfer boundary value problem. *Nonlinear Studies.* **29** (2022), 359–379.

135. A. Salim, M. Benchohra, J. E. Lazreg, J. J. Nieto and Y. Zhou, Nonlocal initial value problem for hybrid generalized Hilfer-type fractional implicit differential equations. *Nonauton. Dyn. Syst.* **8** (2021), 87–100. https://doi.org/10.1515/msds-2020-0127

136. A. Salim, M. Boumaaza and M. Benchohra, Random solutions for mixed fractional differential equations with retarded and advanced arguments. *J. Nonlinear Convex Anal.* **23** (2022), 1361–1375.

137. A. Salim, J. E. Lazreg, B. Ahmad, M. Benchohra and J. J. Nieto, A study on k-generalized ψ-Hilfer derivative operator, *Vietnam J. Math.* (2022). https://doi.org/10.1007/s10013-022-00561-8

138. S.G. Samko, A.A. Kilbas and O.I. Marichev, *Fractional Integrals and Derivatives. Theory and Applications*, Gordon and Breach, Yverdon, 1993.

139. A. M. Samoilenko, N. A. Perestyuk, *Impulsive Differential Equations*, World Scientific, Singapore, 1995.
140. Samina, K. Shah and R. A. Khan, Stability theory to a coupled system of nonlinear fractional hybrid differential equations, *Indian J. Pure Appl. Math.* **51** (2020), 669–687.
141. J. V. Sousa, K. D. Kucche and E. C. Oliveira, On the Ulam-Hyers stabilities of the solutions of Ψ-Hilfer fractional differential equation with abstract Volterra operator. *Comput. Appl. Math.* **37** (2018), no.3, 3672–3690.
142. J. V. Sousa, E.C. de Oliveira, On the Ulam-Hyers-Rassias stability for nonlinear fractional differential equations using the ψ-Hilfer operator. *J. Fixed Point Theory Appl.* **20** (2018), No.3, 96.
143. J. V. da C. Sousa, G. S. F. Frederico, and E. C. de Oliveira, ψ-Hilfer pseudo-fractional operator: new results about fractional calculus, *Comp. Appl. Math.* **39** (2020), p. 254.
144. J. V. da C. Sousa, J. A. T. Machado, and E. C. de Oliveira, The ψ-Hilfer fractional calculus of variable order and its applications, *Comp. Appl. Math.* **39** (2020), p. 296.
145. J. V. da C. Sousa and E. C. de Oliveira, A Gronwall inequality and the Cauchy-type problem by means of ψ-Hilfer operator, *Differ. Equ. Appl.* **11** (2019), 87–106.
146. J. V. da C. Sousa and E. C. de Oliveira, Fractional order pseudoparabolic partial differential equation: Ulam-Hyers stability, *Bull. Braz. Math. Soc.* **50** (2019), 481–496.
147. J. V. da C. Sousa and E. C. de Oliveira, Leibniz type rule: ψ-Hilfer fractional operator, *Communications in Nonlinear Science and Numerical Simulation.* **77** (2019), 305–311.
148. J. V. da C. Sousa and E. C. de Oliveira, On the ψ-Hilfer fractional derivative, *Commun. Nonlinear Sci. Numer. Simul.* **60** (2018), 72–91.
149. J. V. da C. Sousa, M. A. P. Pulido, and E. C. de Oliveira, Existence and Regularity of Weak Solutions for ψ-Hilfer Fractional Boundary Value Problem, *Mediterr. J. Math.* **18** (2021), p. 147.
150. I. Stamova and G. Stamov, *Functional and Impulsive Differential Equations of Fractional Order: Qualitative Analysis and Applications*, CRC Press, 2017.
151. V. E. Tarasov, *Handbook of Fractional Calculus with Applications. Volume 4: Applications in Physics, Part A*, Berlin, Boston, De Gruyter, 2019.
152. S. M. Ulam, *Problems in Modern Mathematics*, Chapter 6, JohnWiley and Sons, New York, USA,1940).
153. S. M. Ulam. *A collection of mathematical problems*. Interscience Publishers, New York, 1968.
154. F. Vaadrager, J. Van Schuppen, *Hybrid Systems, Computation and Control. Lecture Notes in Computer Sciences*, vol. 1569. Springer, New York, 1999.
155. J. Wang and M. Feckan, Periodic solutions and stability of linear evolution equations with noninstantaneous impulses. *Miskolc Math. Notes* **20** (2019), no. 2, 1299–1313.
156. J. Wang, M. Feckan and Y. Zhou, Ulam's type stability of impulsive ordinary differential equations. *J. Math. Anal. Appl.* **395** (2012), 258–264.
157. J. Wang, A. G. Ibrahim and D. O'Regan, Nonemptyness and compactness of the solution set for fractional evolution inclusions with non-instantaneous impulses. *Electron. J. Differential Equations.* **2019** (2019), 1–17.
158. J. Wang and Z. Lin, A class of impulsive nonautonomous differential equations and Ulam-Hyers-Rassias stability. *Math. Methods Appl. Sci.* **38** (2015), 868–880.
159. X. J. Yang, *General Fractional Derivatives: Theory, Methods and Applications*, CRC Press, Boca Raton, 2019.
160. X. J. Yang, F. Gao, and J. Yang, *General fractional derivatives with applications in viscoelasticity*, ed. Walthom: Elsevier, 2020.

161. Y. Zhao, S. Sun, Z. Han, Q. Li, Theory of fractional hybrid differential equations. *Comput. Math. Appl.* **62** (2011), 1312–1324.
162. Y. Zhou, J. R. Wang, L. Zhang, *Basic Theory of Fractional Differential Equations*. Second edition. World Scientific Publishing Co. Pte. Ltd., Hackensack, NJ, 2017.

.

Index

B

Banach space, 3, 4, 6, 7, 9–11, 32, 33, 35, 36, 47, 63, 79, 104, 113, 118, 137, 138, 146

Boundary Value Problem, 2–7, 35–37, 39, 66, 75, 77, 78, 83, 99, 102, 103, 113, 114, 117, 149, 152, 166

C

Caputo, 64, 74, 76

Contraction, 4–6, 8, 32, 33, 44, 51, 60, 70, 87, 94, 110, 124, 130, 143, 144, 156, 159, 160

F

Fixed point, 2, 4–9, 32, 33, 35, 39, 43, 44, 49–51, 56, 57, 60, 68–70, 77, 84, 85, 87, 88, 92, 95, 105, 109, 110, 117, 121, 122, 124, 125, 128–130, 138, 139, 142–144, 153, 154, 156–158, 162

Fractional integral, 4–7, 14, 15, 35, 36, 48, 56, 66, 78, 93, 104, 118, 129, 137, 149, 158

G

Grönwall's lemma, 3, 27

H

Hilfer, 3–7, 15, 16, 35, 36, 48, 55, 56, 63, 66, 75–78, 99, 100, 102–104, 113, 117, 118, 129, 137, 149, 166

I

Implicit fractional differential equations, 4–7, 48, 55, 66, 77, 78, 103, 117, 118, 137, 149

Impulses, 2, 3, 5–7, 77, 78, 103, 117, 118, 137, 146, 149

Initial Value Problem, 2–4, 6, 7, 48, 55, 73, 74, 100, 118, 135, 137, 146

M

Measure of noncompactness, 3–7, 9, 31–33, 35, 77, 117, 148

N

Nonlocal, 3, 6, 7, 129, 130, 135

R

Riemann–Liouville, 16, 18, 36, 53, 74

T

Terminal Value Problem, 74

M. Benchohra et al., *Advanced Topics in Fractional Differential Equations*, Synthesis Lectures on Mathematics & Statistics, https://doi.org/10.1007/978-3-031-26928-8

Printed in the United States
by Baker & Taylor Publisher Services